AF589481

# Redefining Agricultural Libraries in Knowledge Society

## Festschrift Volume in Honour of Dr. Arun Kumar Jain

**Dr. K.Veeranjaneyulu**, M.Com., M.L.I.S.,BGL, PGDLAN, Ph.D., is currently the University Librarian & Professor of Prof.Jayashankar Telangana State Agricultural University, Hyderabad. He has more than two and half decades of professional experience in the field of library and information service. He has participated in more than 100 National Conferences, Workshops and Seminars, and written and edited more than 25 books and contributed around 150 articles. Under his guidance, ten candidates were awarded M. Phil. Degrees and two Ph.D. Currently he is guiding six Ph. D. students. He is a life member of various learned societies, and Chief Editor of the "Indian Journal of Agricultural Librarians & Information Services" (IJALIS). He has been awarded USSHLE-IJILS Param Bhusan B.B.Shukla Millennium Award for excellence in 2007. He received "Certificate of Appreciation" awarded for outstanding contribution in developing KrishiKosh (Agricultural Institutional Repositories Database) in the Libraries of NARES under the e-Granth Project from the Director General, ICAR and National Director, NAIP, IARI, New Delhi. He also received Appreciation Certificate from ICAR for uploading high number of theses i.e. more than 5000 in the Krishikosh during the year 2016.

**Dr. G. Rathinasabapathy** is serving as the University Librarian of Tamil Nadu Veterinary and Animal Sciences University, Chennai. He was awarded M.Com., M.L.I.S., by Annamalai University, M.Phil. (LIS) by Alagappa University and Ph.D. (LIS) by the University of Madras. He has 23 years of professional experience and contributed about 110 research papers to journals, International and National Seminar volumes and participated about 35 International/National level programmes in LIS. He has authored four books in Tamil and three books in English. He is a member in the Editorial Board of Seven Peer Reviewed Journals. He is the President of Tamil Nadu Library Association, Secretary of MALIBNET and life member of 18 professional societies viz., ILA, IASLIC, MALA, AALDI, SALIS, SLA (USA), ISTE, CSI, etc. He has served as the Chairman and Member of about 65 Committees at National and Regional level. He was honoured with many awards including 'Award for Best Paper Presentation' by the Indian Society for Veterinary Pharmacology and Toxicology, "Certificate of Appreciation" by ICAR for his contribution to 'KrishiKosh" Institutional Repository and 'Bharat Gaurav Award' by an International Social Service Association, New Delhi.

**Dr. Rajive Kumar Pateria** is presently working as Deputy Librarian in Chaudhary Charan Singh Haryana Agricultural University, Hisar. He is serving the university library system for last more than 18 years. He did his graduation in Science, Master's degree and Ph.D. in Library & Information Science from Jiwaji University, Gwalior. He has published around 28 articles in prominent journals, books, National and International Conferences. He has edited five books. He is member in Editorial Board of two renowned journals in the field Library & Information Sciences. He has been Co-PI in three national level sub-project named as "KrishiPrabha- An Online Repository of Indian Agricultural Doctoral Dissertations", "Strengthening of Digital Library and Information Management under NARS (e-Granth)" and "Impact Assessment of Usages of Agricultural Information System in National Agricultural Research and Education System (NARES)." His areas of interests are Library Automation, Digital Collection Development & Management and he has vast experience in Library Automation, Digitization, Videoconferencing and other ICT based application in libraries. He visited the USA under NAIP to get hands on experience in digitization and latest LIS techniques.

# Redefining Agricultural Libraries in Knowledge Society

## Festschrift Volume in Honour of Dr.Arun Kumar Jain


— *Editors* —

**K. Veeranjaneyulu**

**G. Rathinasabapathy**

**Rajive Kumar Pateria**



**2017**

**Regency Publications**

***A Division of***

**Astral International Pvt. Ltd.**

New Delhi – 110 002

ISBN 9789389605228 (Int. Edition)

*Publisher's Note:*

*Every possible effort has been made to ensure that the information contained in this book is accurate at the time of going to press, and the publisher and author cannot accept responsibility for any errors or omissions, however caused. No responsibility for loss or damage occasioned to any person acting, or refraining from action, as a result of the material in this publication can be accepted by the editor, the publisher or the author. The Publisher is not associated with any product or vendor mentioned in the book. The contents of this work are intended to further general scientific research, understanding and discussion only. Readers should consult with a specialist where appropriate.*

*Every effort has been made to trace the owners of copyright material used in this book, if any. The author and the publisher will be grateful for any omission brought to their notice for acknowledgement in the future editions of the book.*



***Published by*** : **Regency Publications**
*A Division of*
Astral International Pvt. Ltd.
– ISO 9001:2015 Certified Company –
4736/23, Ansari Road, Darya Ganj
New Delhi-110 002
Ph. 011-43549197, 23278134
E-mail: info@astralint.com
Website: www.astralint.com

***Digitally Printed at*** : **Replika Press Pvt. Ltd.**

**Centre for Poverty Studies &**
**Rural Development**
**Former Director, NAARM , ICAR**
**Hyderabad**
**Telengana State**

**Dr. D. Rama Rao, Ph.D.,**
Director

**Phone: (040) 66534212**
**Mobile: +91 94441273700**

# Foreword

I am very happy to note that a Festschrift volume is being brought out in honour of Dr. Arun Kumar Jain, who has superannuated recently as Head of the Agricultural Knowledge Management Unit, Indian Agricultural Research Institute, New Delhi.

I have known Dr. A.K. Jain for the past thirty years. He joined Agricultural Research Service (ARS) of ICAR in 1976 in the first batch. He has obtained Masters in Physics from IIT, Kanpur and Doctorate in Fibre Physics from IIT, New Delhi. He has held several scientific and responsible position as Assistant Director General (Agricultural Research Information System) at ICAR Head Quarters.

He is an outstanding researcher as evidenced by publications in reputed scientific journals.He is widely recognized for his pioneering efforts in the introduction of ICTs in NARES (ICAR Institutes & SAUs) of India. He has taken proactive steps in creating basic IT infrastructure in NARES through World Bank funded projects - NARP, AHRDP, NATP and NAIP. His contribution in developing internet connectivity, hardware & software, digital content, KVK-ICT Network, e-journals, Common Data Centre and Library Digitization are noteworthy.

His painstaking efforts to network all library and information centres of NARES yielded outstanding results. It emerged as a largest networked library system in India with more visibility. I believe, that is why the Librarians of NARES system are honouring Dr. A.K. Jain with this Festschrift volume.

This volume consists of number of articles contributed by learned scholars from different parts of the country. The volume deals with diversified aspects of Library and Information Science such as ICT in agriculture, knowledge management, e-resource management, information literacy, social media, institutional repositories, RFID technology, library automation, scientometrics and other related aspects. I am sure that readers will find it quite interesting and informative. I congratulate the Editors and others who directly or indirectly responsible for bringing out this Festschrift in honour of Dr.A.K.Jain, who will be ever remembered for his enormous contribution in evolving digital NARES.

**(D. Rama Rao)**

# Preface

It is a great pleasure for us to edit a Festschrift volume in honour of Dr. Arun Kumar Jain, leading scientist, who retired recently as the Head, Agricultural Knowledge Management Unit of Indian Agricultural Research Institute (IARI), New Delhi, contributed enormously for the development of Library and Information System in the National Agricultural Research and Education System (NARES) of this great country.

Dr. A.K. Jain joined ICAR in 1976 through the first Agricultural Research Service (ARS)-Exam in 1976, the first batch. He did M.Sc. in Physics from Christ Church College, Kanpur; worked as research scholar at I.I.T., Kanpur for two years and completed Doctorate in Fiber Physics from I.I.T. Delhi. He worked at different scientific positions, including responsibilities as Assistant Director General (Agricultural Research Information System) at ICAR Head Quarters for five years.

He carried out scientific research in the area of Structure property correlations in fiber, development of electronics instruments for fiber property measurements, Studies on Crop-Weather Interactions etc.

His work is appreciated and supported by about forty publications in reputed journals in addition to presentations in various professional fora. Throughout his professional career Dr.A.K.Jain has held numerous positions in many professional societies.

He has been associated with ICT applications in agriculture sector during last two decades through World Bank funded projects - NARP, NATP, NAIP in creating basic ICT infrastructure all over NARS (ICAR Institutes & SAUs) by implementing internet connectivity, hardware/software and content development strategy, KVK-ICT Network, Data Centre, formulation of ICT component of NAIP and Library Digitization.

Dr. A.K. Jain did a marvelous job in supporting the libraries and Librarians of NARES in their development activities by his unstinted support by all means. As part of 'e-Granth' project, he created a Digital Repository platform "KrishiKosh", a cloud like facility for all institutions under NARS to have their own self-managed repository centrally hosted at IARI, New Delhi. Another major ICT platform developed by Dr.A.K.Jain is Indian Digital Ensemble of Agricultural Libraries (IDEAL), a complete Integrated Library Management System (ILMS) with an Union catalogue "AgriCat" for online access and sharing of library resources of NARS on the concept of Software as a Service (SaaS). Today, the LIS system of NARES is probably the largest and most advanced in the country which would not have been possible without Dr. A.K. Jain. All these merits are more than enough to be honored with a Festschrift.

However, the primary reason for us to organize this Festschrift is that all of us feel deeply indebted for the great support he gave to the LIS community of NARES. When we planned this Festschrift we were well aware of the fact that Festschrif often include a large number of articles on quite heterogeneous topics which stand – if at all – only in loose connection. In the light of the fact that Dr.A.K.Jain was active on so many different fields we decided to refrain from producing a mere collection of articles from different fields of Physics and ICT. Instead, we opted for editing a book on one topic which has been in the centre of Dr.A.K.Jain's interest throughout the last two decades of his service: Redefining Agricultural Libraries in the Knowledge Society.

We would like to use the opportunity to thank Dr.C.Devakumar, Former ADG (EPD), ICAR-Chairman, Dr.P.S.Pandey, ADG (EPHS), ICAR - Co-Chairman; Dr.H.Chandrashekaran, Consultant, National Water Mission, Govt. of India - Co-Chairman; Prof. Prem Singh, Co-Chairman of the Festscrift Committee for their kind guidance and all possible support for bringing out this volume. We are also thankful to the members of the committee and all contributors for their fantastic cooperation.

We would like to thank profusely Dr. Dr.D. Rama Rao, Director, Centre for Poverty Studies & Rural Development and Former Director, NAARM , ICAR, Hyderabad for his Foreword to this Festschrift.

We strongly believe that this volume stands as a small testimony of our respect and admiration for Dr.Arun Kumar Jain, benevolent supporter of Libraries and Librarians of NARES.

**K. Veeranjaneyulu**
**G. Rathinasabapathy**
**Rajive Kumar Pateria**

# About Dr. Arun Kumar Jain

Dr Arun Kumar Jain was born on 22nd July 1954 in a middle class family of Uttar Pradesh, India. He was brought up and educated in Kanpur. After basic education, he joined Christ Church College for his Graduation and Post-Graduation and completed his M.Sc. in Physics in 1974. Thereafter, he was admitted to Indian Institute of Technology (I.I.T.), Kanpur for his Doctorate in Physics. At I.I.T. Kanpur, while working on the problem of theoretical calculations and experimental verification of the energy levels of laser materials, he developed interest in computer science and did several courses in Physics and computer science. The Agricultural Research Service (ARS) attracted him and he join the first batch of ARS in 1976 through an open competition at the age of 22 years. After completing the foundation course with first batch at NAARM, Hyderabad, he was posted at Jute Technological Research Laboratories, Calcutta now known as National Institute of Research on Jute and Allied Fibre Technology (NIRJAFT), Kolkata. Working at NIRJAFT, Dr Jain contributed to research on 'Structure Properties Correlation in Jute and allied fibre' and his interest in computers made him develop micro-processor based instrument for measuring the irregularity in jute yarn and instrument for assessing hairiness of jute yarn. At NIRJAFT, he initiated development of well equipped modern laboratory for physical studies which included procurement and installation of several modern equipments viz., X-ray Diffractometer, Scanning Electron Microscope, Infrared Spectrophotometer, UV-VIS Spectrophotometer and sophisticated fibre testing machines. In 1984, Dr Jain proceeded on Study leave and joined I.I.T. Delhi for completing his doctoral research at Textile Technology Department. His research work focused on 'Structural Property Correlation in heat-set polyethylene terephthalate (PET) fibre' involving in-depth studies on variation in unit cell dimensions using x-ray diffraction techniques, elucidation of morphological ultra-structure to the level of 6 nano-meter using transmission electron microscope, in addition to various other techniques. After obtaining his Ph.D. degree from I.I.T. Delhi in fibre science, he continues his work on structure property correlation in jute and allied fibre at NIRJAFT, Kolkata.

In late eighties and early nineties, as a consequence of development of Personal Computers, the Information and Communication Technology (ICT) was revolutionizing every walk of life by its potential to increase the efficiency and quality of services in every field. Research and Education were amongst first to adopt it. Dr Jain, realizing importance of library in a research institution, took first initiative to introduce the new technology in the library of NIRJAFT by procuring a Personal Computer (PC) and CDS-ISIS software supplies by UNESCO free of cost for managing library functions. This he did in-spite of stiff resistance from the staff due to misplaced perceptions that computers may replace jobs. Implementing CDS-ISIS with the help of librarian, the catalog was converted into electronic form, issue-return records were maintained electronically and periodicals were monitored using the CDS-ISIS 'pascal' programming features of the software. This was appreciated by all library users and paved the way for further computerization.

Meanwhile, in mid-nineties, Indian Council of Agricultural Research (ICAR) took major initiative to harness the power of ICT for strengthening agricultural research and education in country. Under National Agricultural Research Project (NARP-II) standardize set of computer equipments comprising of Servers, computers, UPS, printer, modem etc were supplied to all ICAR Institute, SAUs and zonal research stations with provision for dial-up connectivity. Dr Jain played nodal role at NIRJAFT in establishing dial-up email connectivity using X400 mail from National Informatics Center (NIC), New Delhi, promoted use of computers for day to day work and developed software for accounting, pay slip etc. In 1996, Dr Jain was transferred to ICAR Head Quarters, New Delhi and entrusted with the responsibility of computer application in agriculture, as nodal officer, National Agricultural Technology Project (NATP). During initial phase the priority was to create essential hardware/software infrastructure for harnessing ICT in agriculture research and education, accordingly, standardize set of more advance equipments comprising of Computer Servers, Desktops, UPS, Printer, modem, essential software etc. were supplied, installed and trainings organized at all ICAR institutes, SAUs, Zonal Research stations and Regional Centers. Agricultural Research Information System (ARIS) cells were established at all institutions to host above equipments, coordinate and promote use of ICT in agriculture. During the next phase of NATP, the connectivity was upgraded from dial-up to broadband using a network of VSATs and lease lines installed at 274 locations. Libraries were provided servers, computers and software for automation and electronic catalog. Also, 200 KVKs were connected through VSAT network in an initiative by extension division of ICAR. Dr.Jain played Key role in planning and implementing above goals.

In the year 2002, Dr Jain was given responsibility of Assistant Director General (ADG), ARIS and a new world bank funded project National Agricultural Innovation Project (NAIP) was conceived by ICAR to bring in innovations through technology. Dr. Jain, as ADG(ARIS) played crucial role in planning and drafting of the ICT component of NAIP, deliberately shifting the priority from infrastructure developments to content development for enhancing research and education quality. By this time all institutions were equipped with basic ICT infrastructure and became capable of further expanding and upgrading the ICT infrastructure as per their requirement. Funds were available for ICT through a mandatory policy of government to spend 3-5% of institutional budgets on growth of ICT. Two major policy shift advocated by ADG(ARIS) under NAIP were to fund only content development projects in ICT from different subject matter divisions and the project leadership should be given to subject matter specialists instead of ICT specialists. This was thought important because after the project the content should be owned, updated, further developed and used by the community. This led to desirable results and some very good developments came up in the form of Consortia for e-Resources in Agriculture (CeRA), Rice Knowledge Management Portal (RKMP), Krishi Prabha – a thesis repository, AGROWEB – ICAR institution's standardize portal, eGranth – a vast digital repository (KrishiKosh) of contents from whole NARES and an Union catalogue of ICAR and SAU libraries (AgriCat), centralized data center, ASHOKA super computer for biotechnology applications, Online platform for publication of ICAR Journal, Open Access Policy of ICAR etc. In the year 2008, Dr Jain was transferred at I.A.R.I., New Delhi and played important role as Consortium Principal Investigator (CPI) of e-Granth subproject of NAIP having 38 libraries form allover NARS as partners, aimed at strengthening & digitization of libraries and knowledge management in NARS. These developments have put ICAR and Indian NARES on strong footing for use of modern tools and techniques in ICT for increasing efficiency, improving quality of research output, enhancing education standards and managing knowledge in agriculture sector in effective manner for ultimate benefit of the farmers and society.

Dr Jain's research contributions are evident by about 50 research publications in International and National journals but his more immense contributions in the field ICT application in agriculture research and education will go long way to impact the society.

# Dr. A.K. Jain Festschrift Committee

## CHAIRMAN

### Dr. C. Devakumar

Former Assistant Director General (EP&D)
Indian Council of Agricultural Research, New Delhi – 110 012

## CO-CHAIRMEN

### Dr.P.S. Pandey

Assistant Director General (EP&HS)
Indian Council of Agricultural Research, New Delhi – 110 012

### Dr. H. Chandrasekharan

Consultant, National Water Mission
Ministry of Water Resources, River Development &
Ganga Rejuvenation, Govt.of India, New Delhi – 110 003

### Prof. Prem Singh

President, AALDI
Former University Librarian, CCSHAU, Hisar (Haryana)

## CONVENERS

### Dr. K. Veeranjaneyulu

Professor & University Librarian, PJTSAU, Hyderabad, Telengana

### Dr. G. Rathinasabapathy

University Librarian, TANUVAS, Chennai, Tamil Nadu

## MEMBERS

Dr. Arundhati Kaushik
Mr. C.V. Banker
Dr. D. Chandran
Dr. Chetan Prakash Rajpurohit
Mr. Deepak Dinkar
Mr. Dev Walia
Dr. Devaraj
Dr. R.C. Goyal
Mr. Hans Raj
Mr. H. K. Tripathi
Mr. R.N. Ingale
Dr. U.S. Jadhav
Mrs. Jiji Cyriac
Dr. V.T. Kamble
Dr. S. Khaparde Vaishali
Dr. M. Koteswara Rao
Dr. Madhav Pandey
Dr. R. K. Mahapatra
Dr. A.K. Mishra
Dr. Rajesh Kumar
Dr. Rajive K. Pateria
Mrs. Rajni Bala Grover
Dr. Rakesh Mani Sharma
Prof. L.S. Ramaiah
Dr. B. Ramesh Babu
Dr. D. Ravinder
Mr. S.S. Rawat
Dr. S.M. Rokade
Dr. Seema Parmar
Dr. Shalini Lihitkar
Prof. P.A. Shinde
Mr. B.P. Singh
Mr. T. Sreenivasa Rao
Mr. V. Srinivasa
Dr. Stanley Madan Kumar
Mr. Subhash
Dr. Sudeep Marwaha
Mrs. Usha Khemchandani
Dr. Vaishali Gudadhe (Choukhande)
Dr. Y.Ch. Venkateswarlu

# Contents

# List of Contributors

**Ambhore Sagar Pandit**

Research Scholar, Dept. of Library and Information Science, Dr. Babasaheb Ambedkar Marathwada, University, Aurangabad, Maharashtra State

**Amit Kumar**

Assistant Librarian, M.S.Randhawa Library, Punjab Agricultural University, Ludhiana, Punjab

**Amrender Kumar, Dr.**

Agricultural Knowledge Management Unit, ICAR-Indian Agricultural Research Institute (ICAR-IARI), New Delhi

**Arundhati Kaushik**

University Library, G.B. Pant University of Agriculture and Technology, Pantnagar – 263145, Uttarakhand

**Ashish Sharma**

Agricultural Knowledge Management Unit, ICAR-Indian Agricultural Research Institute (ICAR-IARI), New Delhi

**Bhaskara Rao, P.**

Professor, Dept. of Library & Information Science Dr.B.R.Ambedkar University, Srikakulam, Andra Pradesh

**Chandran, D. Dr.**

Former Professor, Dept. of Library & Information Science, Sri Venkateswara University, Tirupati-517501, Andra Pradesh

**Chandrasekharan, H. Dr.**

Consultant, National Water Mission Ministry of Water Resources, River Development & Ganga Rejuvenation, Government of India, New Delhi – 110 003 chandrasekaranh@gmail.com

**Dev Walia**

Assistant Librarian, University Library CSK Himachal Pradesh Krishi Vishvavidyalaya Palampur -176 062, Himachal Pradesh devwaliacopi@gmail.com

**Jahan Ara, S**

Lecturer in Library and Information Science, SJGC Degree College Kurnool-518 001, Andra Pradesh
jahanaramqbl@gmail.com

**Kamal Batra**

Agricultural Knowledge Management Unit, ICAR-Indian Agricultural Research Institute (ICAR-IARI), New Delhi

**Kopperundevi, S. Dr.**

Assistant Librarian, Veterinary College and Research Institute, Tamil Nadu Veterinary and Animal Sciences University, Tirunelveli – 627 358, Tamil Nadu
devielangowins@gmail.com

**Kumar Kutty, Dr.**

2.Librarian-in-charge, College of Veterinary Science Library, Sri Venkateswara Veterinary University, Proddatur, Andra Pradesh

**Kundan Jha**

Assistant Librarian, Hon'ble Judges Library High Court of Chhattisgarh, Bilaspur 495220, Chhatisgarh.
kundanjha101@gmail.com

**Manoj Mishra**

Assistant Librarian, S 'O' A University, Bhubaneswar, Odisha
mishra_manoj75@rediffmail.com

**Murali Taddi, Dr.**

Course Coordinator and Asst. Professor Department of Library and Information Science Dr. B.R. Ambedkar University, Srikakulam, Andra Pradesh

**Omprakash Sahu**

Computer Application Professional, Pali, District Korba, Chhatisgarh
om70228@gmail.com

**Papegowda, M**

Assistant Librarian, University of Agricultural Sciences, Bangaluru-560 065, Karnataka State
papegowda2009@gmail.com

**Rabindra K Mahapatra, Dr.**

Assoc. Prof., Central University, Tripura
mahapatrark_1962@rediffmail.com

**Raj Kumar Singh, Dr.**

Librarian, College of Horticulture and Forestry, Central Agricultural University, Pasighat, Arunachal Pradesh singhraj1963@gmail.com,

**Rakhee Sharma**

Agricultural Knowledge Management Unit, ICAR-Indian Agricultural Research Institute (ICAR-IARI), New Delhi

**Ramesh Babu, B. Prof.**

(Professor (Retd), University of Madras, Chennai & Former Visiting Professor, Mahasarakham University, Thailand) 22/20B Thangavelu Pillai Garden, First Street, Old Washermen Pet, Chennai 600 021, Tamil Nadu
beerakarameshbabu@gmail.com

**Ramesh, D.B.**

Chief Librarian, S 'O' A University, Bhubaneswar, Odisha
dbramesh@soauniversity.ac.in

**Ranveer Vishakha, B**

Research Scholar, Dept. of Library and Information Science, Dr. Babasaheb Ambedkar Marathwada, University, Aurangabad, Maharashtra State

**Rathinasabapathy, G. Dr.**

University Librarian Tamil Nadu Veterinary and Animal Sciences University, Chennai – 600 051, Tamil Nadu librarian@tanuvas.org.in

**Sanjiv Kapur**

Agricultural Knowledge Management Unit, ICAR-Indian Agricultural Research Institute (ICAR-IARI), New Delhi

**Stanley Madan Kumar, Dr.**

University Librarian (Retd) University of Agricultural Sciences, Dharwad 580 005, Karnataka sunnystanmk@gmail.com

**Thammi Raju, D. Dr.**

Principal Scientist, Education Systems Management Division, ICAR- National Academy of Agricultural Research Management, Rajendranagar, Hyderabad, Telangana

**Uma Jagannath, Dr.**

Assistant Librarian, University of Agricultural Sciences, Bangaluru-560 065, Karnataka State

**Vaishali Khaparde, Prof.**

Professor and Head, Dept. of Library and Information Science, Dr. Babasaheb Ambedkar Marathwada, University, Aurangabad, Maharashtra State khapardevaishali@gmail.com

# 1

# Information and Communication Technology in Agricultural Research

**Dr. H. Chandrasekharan, Ph.D.,**

Consultant, National Water Mission
Ministry of Water Resources, River Development & Ganga Rejuvenation
Government of India, New Delhi – 110 003
*e-mail: chandrasekaranh@gmail.com*

## 1. Introduction

Knowledge, information and data along with the social and physical infrastructures that carry them have been widely recognized as key building blocks for a more sustainable agriculture, effective agricultural science and productive partnerships among the global research community. The processes by which knowledge, information and data are generated and shared are being transformed and reinvented – especially enabled by ongoing developments in the area of information and communication technologies (ICTs) – and that these transformations provide massive opportunities for the entire agricultural research and development community to truly mobilize and apply global scientific knowledge, in ways that are hardly yet appreciated. Catching and successfully harnessing these 'waves' requires strategic investments in capacities, bandwidth, infrastructure, skills, tools and applications, and the adoption of an 'open innovation' mindset that breaks barriers, links data and knowledge, and guarantees the public accessibility of goods generated and captured through science.

## 2. Information and Communication Technology in ICAR

Applications of Information and Communication Technology (ICT) have been so vast today that it is almost impossible to carry out agricultural research without an ICT tool. ICAR has been among the few research institutions of the world wherein Information and Communication Technology development took place in the early 1960s and, as a result an IBM 1620 Model-II Electronic Computer was installed in its one of the institutes i.e., IASRI, New Delhi. About a decade and half later, a third generation computer Burroughs B-4700 system was installed. During 1991-95, the old computers Burroughs B-4700 systems were replaced by new networked PC systems.

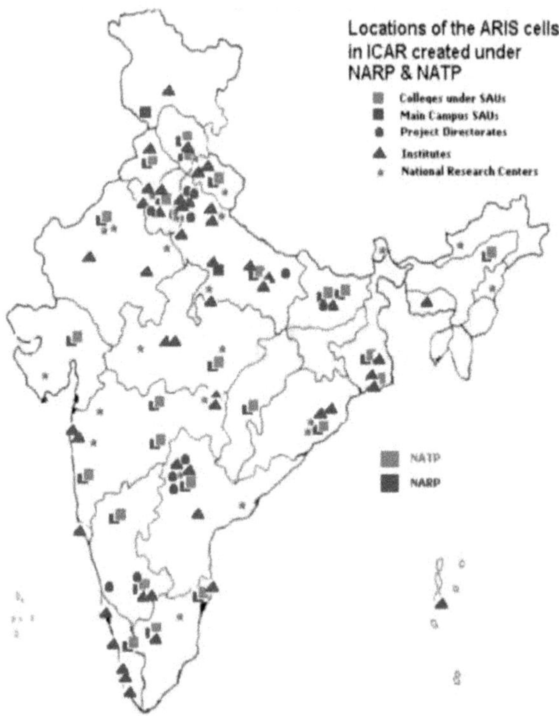


In the eighth five year plan, ICAR embarked upon a project called Agricultural Research Information System (ARIS) to bring the power of information technology to the NARS. Accordingly, Agricultural Research Information System (ARIS) cells were created under the NARP and made functional at 230 locations during 1994–95. In 1998–99, 200 more ARIS Cells were added under National Agricultural Technology Project (NATP).The immense contribution of Dr. A. K. Jain had been among those responsible for the successful implementation of ARIS cells in ICAR. The main objectives of the project were to put information close to the manager and scientist; to improve the capacity of researchers research organizations to organize, store and retrieve information relevant to their mandates and to strengthen national libraries and their access through electronic network, checking the duplication of research and extension projects, dissemination of research findings to the end users, evolving effective information sharing mechanisms among scientists, development agents and farmers. During this phase, a large number of computer hardware and software were installed in various institutes of ICAR and the basic infrastructure required for linking all ICAR institutes was created. The internet and email connectivity were established by linking through dial-up and VSAT using NICNET and ERNET services. The project "Agricultural Research Information System (ARIS)" was implemented to bring information management culture to National Agricultural Research System (NARS) so that agricultural scientist can carry out research more effectively by having systematic access to research information available in India as well as in other countries, better project management of agricultural research, and modernization of the office tools. ARISNET has four information modules namely Agricultural Research Personnel Information System (ARPIS); Agricultural Research Financial Information System (ARFIS); Agricultural Research Library Information System (ARLIS) and Agricultural Research Management Information System (ARMIS).Some of the software developments during NATP and at ARIS cells of Institutes were:

## Database Management Systems

- For Gene bank management by NBPGR, New Delhi
- Project Information and Monitoring System-PIMS for NATP, IASRI, New Delhi
- Identification & Management of Nematodes in India, NCIPM, New Delhi
- RAINSIM for Rain water simulation by CSSRI, Karnal

## Information Systems

- PDGRIS for Poultry Disease Diagnostics & Remedy, CARI, Izatnagar
- AGRI-IS on Animal Genetic Resources of India, NBAGR, Karnal
- Agricultural Pest Information System by CTRI, Rajamundry
- Pulse Information System for UP by IIPR, Kanpur

## Data CDs

- Pulse Variety CD by IIPR, Kanpur
- Potato pests CD by CPRI, Shimla
- CD on machines developed under FIM by CIAE, Bhopal
- CDs on Workshop on Bio-informatics and statistics and
- CIFA at a glance by CIFA, Bhubaneswar

## Application Software

- S/W to implement HACCP seafood processing plants, CIFT, Cochin
- Identification of Eggs & Larvae of Parasites, ICARRC for NEH Region, Umiam, Meghalaya
- Library Information System package by NRC for Banana, Trichy

## Expert Systems

- Grape, Cabbage, Mushroom Cultivation Expert Systems by IIHR, Bangalore
- Cotton Insect Pest Management System by CICR, Nagpur
- Statistical Quality Control for Dairy Plants by NDRI, Karnal
- Databases on Inventory of camel herd, Biometry, Breeding tract, Reproduction of camel herd by NRC on Camel, Bikaner.

## Data warehouse

- Central data warehouse entitled Integrated National Agricultural Resources Information System (INARIS) was developed by IASRI, New Delhi-110012.

# 3. ICT Initiatives under NAIP

3.1. **Consortium of e-Resources in Agriculture (CeRA):** The NAIP has established a Consortium for e-Resources in Agriculture (CeRA) for providing online access to e-journals and resources in over 150 libraries. This will greatly help the resource-poor SAUs who find it difficult to subscribe to exorbitant international scientific journals. This will provide a new and competitive research environment where the scientists will have greater access to quality research material and spend less time in literature search. Increased scientific output, quality publications in high impact journals, greater visibility of research by our scientists at the international level are some of the positive impacts envisioned.

3.2. **Strengthening Statistical Computing for NARS:** To strengthen the high end statistical computing environment and to sensitize for the scientists in NARS with the statistical computing capabilities available for enhancing their computing and research analytics skills the project has been launched. Under this project all ICAR institutes have been provided SAS package for statistical data analysis. This has given a major impact on research quality.

3.3. **Establishment of National Agricultural Bioinformatics Grid (NABG) in ICAR:** To bridge the gap between genomic information and knowledge, utilizing statistical and computational sciences, the network of institutions based on two-tier architecture in bioinformatics has been initiated. It opens up new vistas for downstream research in bioinformatics ranging from modeling of cellular function, genetic networks, metabolic pathways, validation of drug targets to understand gene function and culminating in the development of improved varieties and breeds for enhancing agricultural productivity. NABG helps in capacity building for research and development in agricultural bioinformatics and in turn agricultural biotechnology and provides platform for inter-disciplinary research in cross-species genomics. It is observed that information and knowledge generated through research on bioinformatics from the genomic knowledge base and experimentations in different sectors of agriculture are able to evolve internationally superior competitive varieties/breeds and commodities in agriculture.

3.4. **KrishiPrabha Indian agricultural dissertations repository:** Doctoral dissertations - the end product of doctoral research is a rich source of original information, access to which is severely restricted due to absence of information in digital form. NAIP is funding a project wherein over 10,000 doctoral dissertations produced in India during 2000-2006 have been digitized and made available online and offline. NAIP would also promote appropriate policy and create a national repository of these resources. Currently KrishiPrabha forms a part of E-Granth project.

3.5. **Agroweb - Digital Dissemination System for Indian Agricultural Research (ADDSIAR):** For enhancing the effectiveness of ICAR's web-based dissemination and publishing platform by exploiting new generation Web2.0 technologies, this project has been initiated to address issues like visual and domain name uniformity, content quality, web applications for e-commerce (tenders) and e-publications (newsletters, research reports and papers), meta data standards, RSS feeds, discussion forand tools for hosting information on ICAR policies, regulations, technologies, opportunities, events, etc. The uniform guidelines developed and implemented in the project at selected institutes on pilot basis shall be scaled up to cover all ICAR institutes.

A web portal has also been created and maintained at ICAR headquarters, New Delhi, which acts as a common gateway to all ICAR institutes as well as to Agricultural Research and Education System in the country.

3.6. **Development of e-Courses:** Indian National Agricultural Research System along with the State Agricultural Universities caters to the human resource needs of more than 60,000 agricultural graduates and scientists every year. The traditional methods of educating this diverse mass in diverse geographical areas are difficult in the changed context of globalisation and competitiveness. E-Learning is the current trend in synchronous classroom education as well as asynchronous distance education. The NAIP supported projects to develop e-courses for five undergraduate programs in Agriculture, Veterinary Science, Horticulture, Fisheries Science, Home Science and Agricultural Engineering. It amounts to developing over 300 courses in e-format and making them available both online and offline through CDs and DVDs. This exercises benefitting students and teachers alike in making available uniform and quality learning material in all the universities and colleges throughout the country.

3.7. **ICT-mediated Knowledge Management and Dissemination:** A key step in the re-vitalisation of the extension system identified by the policy makers is to enhance knowledge flows between different actors and stakeholders in the extension process. Contemporary developments in ICT are accepted by the policy makers as providing new opportunities to make it happen. A consortium of institutions of excellence in ICT research,

SAU's and agricultural research organisations have come together to build a comprehensive and integrated set of processes and platforms to support and promote knowledge flows and exchanges between different stakeholders. This shall result in a model knowledge organisation linking research and education sectors with the upcoming and revived extension processes in Indian agriculture. The ICT resource partners will develop a process for advanced content organisation in Indian agriculture called Agropedia (www.dealindia.org), which is based on the successful model of Wikipedia. They will also deploy an advanced platform called AAQUA (http://aaqua.persistent.co.in/aaqua/forum/index) to facilitate interactive discussions between experts and farmers (as well as within the community of experts) that will work in both online and offline modes.

3.8. **Centralised Data Centre (e-Governance):** ICAR is in the process of introducing Enterprise Resource Planning (ERP) systems, financial management system, portal applications, e-learning and knowledge management modules and messaging servers for effective functioning of the Council. To run all these applications, a Data Centre is being developed to house computer systems with security mechanisms and associated components, such as telecommunications and storage systems. This shall help in redundant or backup power supplies, redundant data communication connections, environmental controls (air conditioning, fire suppression, etc.), and special security devices for 24 x 7 availability. This is expected to include 10,000 user nodes. This ICAR-NET shall be strengthened through creation of secured and scalable intranet with centralised data centre. The data centre would be located at ICAR Headquarters, New Delhi with database servers, web server, video server, application server, unified mail servers, messaging server and antivirus gateway. A backup/disaster recovery site shall also be created in subsequent phases at geographically different location to provide full data centre redundancy for uninterrupted and seamless access to the resources. The complete infrastructure shall be managed by the ERP software. The data centre shall address all the key requirements including the servers, storage area network, physical infrastructure, power conditioning devices, security, messaging and enterprises management system. The IP based video conferencing (IPVC) and IP Telephony systems will be extended to each and every institute of ICAR and State Agricultural Universities (SAUs) for effective communication.

3.9. **On-Line Management Information System (MIS) and Financial Management System (FMS):** A need was felt to have a financial and accounting system which would not only meet the present day accounting requirements but also help the day-to-day monitoring of quantitative and qualitative aspects from operational as well as management point of view in timely manner. There was also a need for a secure, robust, flexible and scalable FMS/MIS solution that provides a platform for scaling up department's growth, diversification, and expansion. Therefore, a web-based real time online Financial Management System has been started.

This project envisages the plan for developing and rolling out FMS at ICAR and its institutes in the initial phase, followed by MIS implementation plan which includes Project Management, Inventory Management the HRMS (Human Resource Management System) and payroll package. The Management Information System is expected to enable a meaningful extrapolation, forecasting and projections. Sound and updated information would serve as a base for effective managerial control and timely decision making.

Advantages of the system are

- System report generation: Requisite reports (SOE, Balance Sheet, Income Expenditure report, Annual Receipt Payment Report) can be generated directly from the system. In addition a large number of reports customized as per the institute's need can also be generated.
- Real-time centralized database with online data feeding: Since all the data is fed online into a central database, it ensures a real-time availability of data at any point of time which is critical for management review purposes.
- Expenditure review, financial planning and decision making: Since all data is available online this system would ensure quick decision making by the management. It would also help in effective monitoring of cash flow and the utilization of funds under various expenditure heads across all the institutes.
- Transparency and removal of discrepancies: All the modules in the system are synchronized to work in tandem leaving no scope for discrepancies in the data. This would ensure complete transparency while preparing the reports from the data fed into the system.

The system looks very easy and simple however, the application of this system requires a lot of understanding and this utilizes a huge internet bandwidth which is a major constraint in several institutes of ICAR.

## 4. Issues and Challenges

The use of internet has received a paramount importance in all walks of life and has its own importance similar to water and electricity. The use and applications of internet has increased many folds. There are several challenges for the high speed internet connectivity and one of them is bandwidth management.Internet security has become part of everyday life where security problems impact practical aspects of our lives. Even though there is a considerable corpus of knowledge on tools and techniques to protect networks, information about what are the actual vulnerabilities and how they are exploited is not generally available. This situation hampers the effectiveness of security research and practice. Understanding the details of network attacks is a prerequisite for the design and implementation of secure systems.

Some of the following important cyber security challenges attract our attention.

- *Verifying User Identity*: How can others know it's you? Communication is approaching near continuous between friends, family, businesses & services. With current authentication standards, often we take on faith that we're being contacted by the "real" sender the message claims. It's one thing if the imposter is just sending e-mails, but what if it's your bank or retirement account that doesn't know it's not you?
- *Preventing User System Hijacking*: Even with better and better firewalls and anti-malware software for users, malicious programs (like viruses, worms or trojans) that take control of a user's computer and programs are an ever-present threat. Once the malicious program has control it can wreak havoc acting as the user, attacking other contacts while masquerading as the hapless victim.
- *Protecting User Confidential Data*: More and more services are moving to the Internet. Interoperation between the various services is becoming more frequent and more complex. Financial transactions from sales to investments online are becoming ubiquitous. The risk of sensitive & high-value data exposure and criminal access to that data increases all the time.
- *Securing Web Applications*: Developers and application providers want their applications to be available quickly and easily to anyone in the world, from any platform from a phone to a kiosk. Having users hassle with anything more than a simple password seems too much to ask. It needs to at least consider the option for certificates, multi-factor authentication, multi-stage authentication and so forth.

## 5. Concluding Remarks

Contents of this article are taken from the report on 'ICT in Agriculture – Roadmap' and the author was a member of the Committee that prepared the report. Important suggestions are:

- Policies for bandwidth management is needed to restrict how bandwidth is used and policy development must be consultative, supported and led from the top.
- Digital library needs a compulsory participation of the institute in the digital revolution. In this regard, efforts made by Dr. A. K. Jain are praiseworthy.
- ICAR can have a partnership with National Resource Centre for Free and Open Source Software (NRCFOSS) and develop FOSS software for use in Agriculture Research and Development.
- ICAR should mandate establishment of Institutional Repositories in all its institutes for world-wide electronic distribution of the peer-reviewed journal literature and completely free and unrestricted access to it by all scientists, scholars, teachers and students will accelerate research, enrich education. This is partially achieved by institutions associated with E-Granth project.

- ICAR Data Centre be strengthened to host Cloud Computing facilities for the Website, Educational Courseware, Repositories and Journal hosting using FOSS.
- As and when an issue on Indian Agricultural Research Activities is raised/ reported in any media, the concerned institute(s) should immediately respond through proper documents (report, article, leaflet, etc.) and send a copy of the same to ICAR and, if need be, reported at the appropriate forum.

## Few words on Dr. A. K. Jain:

Dr. A. K. Jain is among the few scientists in ICAR, who initiated ICT in Agriculture. He is from the first batch of ARS in 1976 in the discipline of Physics. I, also from the discipline of Physics from the 3rd batch in 1978, fondly recall the first meeting with Dr. Jain, at JTRL, Calcutta (now, NIRJAFT, Kolkatta) way back in 1978. Subsequently during my stay at NRL, interacted with Dr. Jain along with Dr. Rama Rao (currently Director, NAARM) when both of them were doing their Ph.D. Programme in IIT, Delhi. Thereafter, my association with Dr. Jain grew in leaps and bounds during the implementation of NAIP, in which, with his kind support, USI, with just three scientists was, sanctioned four sub-projects of NAIP, as CPI/CCPI. During 2009-10, the then Director, IARI informed me that Dr. A. K. Jain was to join IARI on transfer from Kolkatta and would like him to be posted in Division of Agricultural Physics instead of USI (although USI was better suited for Dr. Jain) for the simple reason that, had Dr. Jain was posted at USI, he would be In-charge of the Unit and that my status would be disturbed. While conveying this to Dr. Jain, I informed him that it would be my pleasure to work under his supervision and requested him to join USI. Dr. Jain, to my surprise, politely rejected my request but informed that he was always available even after joining Division of Agricultural Physics. It is in the fitness of things that Dr. Jain was made to look after USI (now AKMU) soon after my superannuation in May 2014. All along, it was my pleasure to interact with Dr. Jain and I always admire his way of attending to official responsibilities with success. This article is presented in honour of Dr. A. K. Jain, on his superannuation. I wish him and his family a long and healthy life.

# 2

# Knowledge Management and Library Profession: A Symbiotic Relationship

**Dr. Stanley Madan Kumar**

University Librarian (Retd)
University of Agricultural Sciences
Dharwad 580 005
e-mail : sunnystanmk@gmail.com

**ABSTRACT**

*Knowledge Management (KM) is a strategy of getting the right knowledge to the right people at the right time and helping people to share and put information into action. Technological developments lead to changes in work and changes in the organization of work. As a result, required competencies are also changing. With the perpetual advances in digital technology and ICT, library has been incessantly evolving. Information professionals have evolved from traditional cataloguer and reference service providers to value-added service providers-- the Knowledge Managers. To meet this demand, library science education requires effective preparation of new generation librarians to effectively use the innovative technologies in their professional practices This paper is an attempt to briefly describe the role of information professionals in KM and the challenges unique to 21st century.*

***Key words:*** *Digital Libraries, KM, LIS Schools, Information professionals*

## 1. Introduction

Multidisciplinary knowledge management is one of the hottest topic of discussion in modern times. Today, more and more information is being stored digitally and disseminated electronically and all types of materials are available in CD-ROM and online.

## 2. What is Knowledge Management?

Knowledge Management (KM) is a systematic approach that ranges from technology-driven method of accessing, controlling and delivering information. KM is not only about managing knowledge assets but managing the process including developing knowledge, preserving knowledge, using knowledge and sharing knowledge (Vinitha & Mynoon, 2005).

KM is a strategy of getting the right knowledge to the right people at the right time and helping people to share and put information into action in ways that strive to improve the organization's performance (Perera).

Since 1990s the concept of KM has been one of the chief focus of study and debate in all fields of study and more so in library science and information science. Many authors are skeptical whether KM is the same old librarianship under 'new brand name'.

## 3. Transformation of Information

Every organization generates huge quantity of data and information. Organizations need to build up and continually replenish their capabilities or else they will be naturally weeded out in the current era of globalization where knowledge is the most important sset for survival. Without effective management it is difficult to identify and locate the information required in a specific situation. In order to overcome this hurdle, KM came as a strategy to develop organizational knowledge and capacity to derive relevant knowledge from information. That is the reason why knowledge is portrayed as the 'transformation of information,. Information is a building block of knowledge which is the refined relevant and actionable aspect of information (Jain,2009).

## 4. Higher Education and KM

Higher education is becoming increasingly dependent on the intellectual capital of staff and their capacity to grow and survive in a dynamic environment. According to Alabi (2004) post-graduate education is knowledge acquisition, knowledge incubation, knowledge amplification, and knowledge dissemination. It is self evident that information is a key resource which permeates post-graduate teaching, learning, research and publishing.

The emphasis on the university's role as 'provider' of continuous professional development has now changed to meet the needs of the employees who are also 'learners' as well as employers who are also 'developers and users' of high level knowledge incorporated and generated by the work-based programmes (Garnett,2001).

The emergence of knowledge-based society is changing the global economy and the status of education. The quantity of information is exponentially greater than that was available a few years ago and this rate of growth is accelerating (Unesco.1998).

## 5. Need for In-service Training Programmes

New trends and developments in digital technology and ICT worldwide pose great challenges for library professionals especially in developing countries. The irony of the situation is that the LIS educators trying to teach 21st century skills which they have not in their turn had the chance to practice in an organizational environment. These factors emphasize the need for in-service training programmes for both teaching and practicing librarians. In this way they can jointly add value to the parent organization and create a cycle of continuous professional development. Inadequate infrastructure, poor human and financial infrastructure, lack of faster access to necessary information and resources for earning are some of the major issues library schools are to contend in developing countries.

One way of transforming the teaching-learning process is to properly harness and integrate the efficacy of ICT into library science education and training.

## 6. Necessity of Need-based work-force

Information society demands a work-force that can use technology as a tool to increase productivity and creativity. Specific jobs and job markets require certain sets of unique competencies. This involves identifying reliable sources of information, effectively accessing them, synthesizing and communicating that information to colleagues and associates (Hawkins, 1998).

There is an increased awareness among LIS professionals of their potential contribution to KM with a high level of agreement on its positive implications for both individuals and the profession. Courses on KM are being included in the curricula of LIS schools. If 21st century skills are more commonly taught in LIS courses it would help bridge the gap between skills acquired in class rooms and those required by labour markets (Ferreira et al, 2007).

LIS courses must be designed to develop generic and specific skills in the future librarians and prepare them for prevailing labour market.

## 7. Importance of Need-based Curriculum

Need-based curriculum needs to be conceived in relation to market needs and employer perceptions about the competencies of professionals (Rehman, 2008). An appropriate hybrid of teaching and learning methods would provide meaningful learning environment in library schools This would be achieved by steady connection to the information super-highway which would enhance LIS schools' capacity to respond to the new challenges in information processing and dissemination (Igwe 2005).

Providing opportunities to meet the basic learning needs of information professionals is the first step toward preparing library schools for emerging global society.

For KM to be successful, it is necessary to understand how people in organizations learn, how they implement what they learn and how they share their knowledge. Higher education has been concerned about creating a learning environment in which staff and learners learn their experience in teaching and learning, coupled with exposure to research (Rowley,1999).

## 8. Role of Higher Education Institutions in KM

For higher education institutions knowledge is power because the employability of individuals is directly related to their knowledge. Higher education concern itself with knowledge and hence knowledge creation, dissemination and learning are regarded as core activities. In area of knowledge access and knowledge repositories progress in higher education institutions has taken place but not enough. To excel in future, they have to manage explicitly, systematically and comprehensively.

Top management of the university plays crucial role in managing the intellectual asset. The important role of KM should first be understood by top management before necessary decisions can be made. They need to understand the economic consequences of KM practice in higher education institution.

Management style need to change in order to recognize the growing trends to manage knowledge workers differently from normal traditional management practices. KM could help to maximize the potential of knowledge workers in order to increase their productivity and output of skills and intellectual capability.

## 9. Librarian and KM

There seems to be a gap in KM perception by librarians. Some focus on information professionals' new roles, some of skills, some at the challenges faced by conventional librarians. There is an absence of holistic approach.

The new media has not only added value to the interactive communication, also provides powerful new means of accessing information to support teaching, learning and research programmes.

Generally, the term 'information professional' is used to denote 'modern librarians'. With the advances achieved by the ICT, anyone who deals in the provision of information at any stage, whether in terms of technology , education, systems or actual information services, can be called 'information professional'. However, this paper refers to academic librarians and archivists who are revamping to become Knowledge Managers and LIS Educators (Abels et al, 2003).

The thrust of KM in relation to information professionals is to enhance accessibility of information to promote strong relationship with clientele and relevant stakeholders by strengthening knowledge flow, offering value adding services and training cost and time effectively, customized to organizational needs (Jain ---??). Due to information overload, information professionals are needed more than ever to quality filter and provide required information in an usable form.

## 10. National Knowledge Commission of India

The National knowledge Commission (NKC) was set up to guide and advice for the proper utilization of the vast knowledge base of India toward building an equitable knowledge society. Library and Information systems and services is one of the major areas of Access to Knowledge. With the mandate of NKC, different types of libraries have been networked through resources.

There is need to create mechanisms and institutions which will bring about a paradigm shift in the LIS scenario, to bridge the gap between the information poor and the information rich. Libraries of IIs and IIMs have taken up the right kind of initiatives in digitizing resources, creating institutional repositories and developing web sites. Several universities libraries have started to revamp and re-engineer their services through electronic media by e-subscriptions, e-journal consortiums and priced online databases (Das Gupta,2007).

## 11. How Library Professionals can contribute to KM?

Main goal of librarians should be to ensure that their clientele know what information resources are available to them and how the library staff can facilitate access to them. Information professionals need to understand the role of knowledge in every area of the organization. They have to involve themselves in the issues generally not considered to be under their purview (Jain, 2009)

- Empowering knowledge workers by developing necessary tools and services , facilitate direct access to knowledge 24/7, any place.
- Bridging the gap between skills acquired and job market. Presently there is considerable gap between class room lectures and professional practice need in the job market.
- Leading in information management skills. Develop good rapport with the executive management.
- Innovative teaching for greater collaborative and closer interaction.—Using multimedia resources provide web-based online education opportunities. There is so much information and knowledge and collaboration effort is critical to prosper in the present day Information Age.
- Creating KM awareness. Often there can be resistance to new ideas and initiatives from the information professionals themselves. Create awareness of the benefits of KM among students, faculty and other stake holders, particularly LIS discipline.
- Leading in ethical and legal issues.—Device and implement standards for professional ethics towards information usage, privacy, plagiarism, copyright etc.

## 12. Conclusion

KM has several benefits. It increases the amount of learning that is taking place; makes the work less frustrating; makes the learning organization a reality and creates knowledge insight and understanding that can assist people in their daily lives outside the work (Bassi et al,1998).

Successful KM implementation depends upon the integration of many different aspects or an organization. Proper planning and continuous evaluation are needed to ensure that all aspects of KM are being implemented effectively. The skill requirements for a knowledge worker could vary drastically depending on his specific areas of responsibility.

Library professionals need to focus on their user community's knowledge and remain updated in their professional knowledge in order to provide value-added services to their clientele, anywhere, 24/7 and in desired format.

With the ever growing electronic availability of information on both national and global networks, may academic libraries have turned their attention to providing access rather than building traditional collections.

The more knowledge is shared, the more valuable it becomes. Knowledge comes as a person uses information and combines it with their personal experiences. In order to share knowledge the individual should possess the knowledge or at least where to obtain the required knowledge. KM is no more an optional extra for 21st century information professionals but a mandatory discipline if they want to survive in the digital era.

## References

Abels (E), Jones , Lotham (J), Magnoni (D) and Marshal (JC), 2003. Competencies for information professionals of the 21st century. http://wwwsla.org/pets/competencies

Alabi, 2004. Evolving role of ICT in teaching, research and publishing. Nigerian Tribune (Friday)(30 April):30–31

Bassi (L), Cheney (S) and Lewis (E), 1998. Trends in work place learning: Supply and demand in interesting times. Training & Development 52(11):51–54, 62–24, 66–75

Broadbent (M), 1998. The phenomenon of KM: What does it mean to the information profession? Information Outlook 2(5): 23–36

Chase (RL), 1998. Knowledge navigators. http://www.sla.pubs/serial/io/1998/sep98/chase/html

Das Gupta (K), 2007. The Knowledge Commission and Libraries: A vision for the future. Delhi Library Assn. Networking Professionals, Dr. Ranganathan Lecture Series (Fifth Lecture). http://www.dlaindia.org/srr7/html

Du Toit (ASA),2000. Knowledge Management is indispensable component of the strategic plan of South African universities. South African Journal of Education 20(3):187-191

Ferreira (F), Santos (JN), Nasimento (L), Andrade (RS), Barros (S)m Borges (J), Silva (HP) and Jamberia (O), 2007. Information professionals in Brazil: Core competencies and professional development. Information Research 12(2): Paper 299.

Garnett (J),2001. Work-based learning and the intellectual capital of universities and employers. The Learning Organization 8(2):78–82

Harrison (RT) and Leitch (CM),2000. Learning and organization in the knowledge-based information economy: Initial findings from a participatory research case study. British Journal of Management 11: 103–119

Hawkins (RJ), 1998. Ten lessons for ICT and Education in the developing world: IN World Bank Development Indicators. World Bank, NY :Pp 38–43

Igwe (UO), 2005. Harnessing information technology for the 21st century Library Education in Nigeria. Library Philosophy & Practice 7 (1) (Spring)

Jain (Priti), 2009. Knowledge Management for 21st century information professionals. Journal of Knowledge Management Practice 10 (3)

Perera (Kamani). Changing roles of library professionals in the Knowledge Society. www.academia.edu

Rehman (SU), 2008. Analysing corporate job market for developing information and knowledge professionals: The case of Developing Nations. Malaysian Journal of Library & Information Science 13 (1): 45–58

Rowley (J), 1999. What is Knowledge Management? Library Management 20 (8): 416–419

Unesco, 1998. World Education Report: Teachers and teaching in a changing world.

Vinitha (K) and Mynoon (M), 2005. Knowledge Management: Emerging challenges for LIS professionals. Proceedings of the National Convention on Library & Information Networking (NACLIN-2005): 122–135

Wilson (T), 2002. The nonsense of Knowledge Management. Information Research 8(1): 39

# 3

# Open Access Movement as a Means to Innovation in Research: Boon or Bane?

**Prof. B Ramesh Babu**

*(Professor (Retd), University of Madras, Chennai & Former Visiting Professor, Mahasarakham University, Thailand)*
22/20B Thangavelu Pillai Garden, First Street, Old Washermen Pet, Chennai 600 021
email: *beerakarameshbabu@gmail.com*

## 1. Introduction

*"The Open Access Movement has fought valiantly to ensure that scientists do not sign their copyrights away but instead ensure their work is published on the Internet, under terms that allow anyone to access it". --- Aaron Swartz*

Scholarly communication is the process of academics, scholars and researchers sharing and publishing their research findings so that they are available to the wider academic community and beyond. It encourages the advancement of effective, extensible, sustainable, and economically viable models of scholarly communication that provide barrier-free access to quality information. It is the life-blood of the university's teaching and research mission. With the advent of new technologies, the nature of scholarship and scholarly communication has expanded beyond traditional print formats to include other means of dissemination: email, pre-print servers, e-journals, e-books, e-reserves, distance learning, etc. In an online environment, issues of copyright, intellectual property rights, and the long-term preservation of digital assets are posing new challenges to faculty, administrators, and librarians (Source: https://www.lib.uci.edu/what-scholarly-communication). The Open Access (OA) Movement is a plan to make the IT revolution a reality in the world of scholarly communication. The Open Access Initiative (OAI) in the beginning of new Millennium has offered new vistas and opened an era of open access to information. In the recent years much discussions and initiatives are taken in the area of open access. Open access, a philosophy facilitates availability and distribution of scholarly communication freely, as a means to solve the problem of inaccessibility primarily due to financial constraint particularly in the context of developing countries (Ghosh and Das, 2006). India has made an important contribution towards the growth of Open access publishing. Not only governmental funding agencies but also learned societies, associations and publishers have taken a step towards Open Access Movement in a right direction.

## 2. Concept of Open Access

"Open access" refers to the practice of making scholarly research available online for free upon publication (or soon after). Implemented by academics, institutions, journals, and major funding bodies, open access policies allow everyone across the globe to benefit from the latest findings and discoveries. "Open Access (OA) literature is digital, online, free of charge and free of most copyright and licensing restrictions. What make it possible are the internet and the consent of the author or copyright-holder" (Suber, 2004). He further states that open access contents are not restricted only to peer-reviewed research articles, they can be in any formats from texts and data to software, audio, video, and multi-media (Suber ,2010).

According to BOAI (Budapest Open Access Initiative), the concept of Open Access refers to "[the] free availability on the public internet, permitting any users to read, download, copy, distribute, print, search, or link to the full texts of these articles, crawl them for indexing, pass them as data to software, or use them for any other lawful purpose, without financial, legal, or technical barriers other than those inseparable from gaining access to the internet itself" (BOAI, 2002). The Guru of Open Access Harnad (2008) has described the characteristics of Open Access as, "Information, which is free, immediate, permanent, full-text, on-line and accessible". Harnad further suggests three main justifications of OA , namely, "to maximise the uptake, usage, applications and impact of the research output of your university; to measure and reward the uptake, usage, applications and impact of the research output of your university (research metrics) and, to collect, manage and showcase a permanent record of the research output and impact of your university".

Open access can take two paths. One is publishing in open access journals, which is called 'gold open access' and the other is depositing in a digital repository which is known as 'green open access'. Digital archives or repository is a set of systems and services which facilitates ingest, storage and retrieval management, display and reuse of digital objects. The rationale behind the open source philosophy is simple collaborative efforts and co-creation by different people sharing their individual knowledge enhances progress and facilitates development of better software. Further it can be stated that these are online resources to scientific and scholarly research literature. Open source is one tool of many needed in overcoming the digital divide. It enables developing countries like India to take part in research, the Internet and other forms of information or knowledge usage which is done through the computer. Moreover, open source enables the use of programs that help people to improve their work and make the life easier. The advantages of open source are obviously present, but they may be overshadowed by the disadvantages. With regard to developing countries one problem is the language and the ability to learn or better the possibility to learn certain things.

Wilinsky (2003) identified nine flavors of open access. They are:

1. E-print archive (authors self-archive pre – or post prints)
2. Unqualified (immediate and full open access publication of a journal)
3. Dual mode (both print subscription and open access versions of a journal are offered)
4. Delayed open access (open access is available after a certain period of time)
5. Author fee (authors pay a fee to support open access)
6. Partial open access (some articles from a journal are available via open access)
7. Per capita (open access is made available to countries based on per-capita income)
8. Abstract (open access available to table of contents/abstracts) and
9. Co-op (institutional members support open access journals)

## 3. Open Access Movement

The OA movement was started by Michael Halt with Gutenberg Project in 1976. In the year 1989, the first Free Online Journal Psychology by Stevens Harnard hosted by the University of Southampton was launched. Subsequently, the Project Muse was launched in the year 1993. On February 15, 2007 the National day for OA was celebrated all over the America for the first time. A little later, during 19–23 October 2011, OA week was celebrated all around the world. However, in India, the OA movement was began late with the project Digital Library of India (DLI) in 2002, with an aim to digitally preserve the significant works of scientific, artistic and literary nature. Open access is a movement whose major objective is to make available peer review research available on the Internet without any hassles of restriction on its access and affordability, thus bridging digital divide between developing and developed world. The major objective of open access movement is to deal with 'crisis' in the present scholarly communication. The goal of the movement is to ensure the free, equitable, organized and wide distribution of the output of scholarship and research on the public Internet (Suber, 2003).

## 4. Need for Open Access Movement

Developing countries like India, it is very much necessary for accessing free access scholarly journals. The trend has been changed for publishing to Open Access Movement.

- Most reputed journals are published in developing countries (US/Europe) and are monopolized by few publishers
- Purchases are in foreign currency
- Even best indigenous research is published in foreign journals

## 5. OA Movement is a Boon for Academic world

For the developing world, open access movement has come as a boon. Open access is an alternative business model for the publication of scholarly journals. It makes articles freely available to readers on the Internet and covers the costs associated with publication through means other than subscriptions. Open access publishing model fosters growth of research and development activities in academic and research institutes. The open access initiatives facilitate the free availability and distribution of scholarly communication. For the developing world they offer a means to solve the problem of inaccessibility of scientific information, which was primarily due to financial constraints.

There is no doubt that open access is a significant contributor to innovation and, growth and development. It has the potential to deliver critical information to change the lives of people. Open access provides a new paradigm for libraries and librarians. It provides unprecedented opportunities for librarians to deliver rich content in the quest to ensure that there is growth and development. It has the potential to deliver opportunities to loosen the noose of those communities that are strangled because of the lack of critical information (Tise, 2010).

The open source movement has produced vast quantities of valuable software, as well as raised public awareness to issues like open access and open content. Digital libraries, open access and Open source software are a natural outgrowth of the open models of exchange that help societies grow and prosper. Benefits of open access, open sources and digital libraries are numerous and the use of open source software in the library and information community is growing. The key term to connect digital libraries, the open access movement and open source software is 'open'. In fact, digital libraries are sometimes referred to open digital libraries, and open models, such as open archives, have emerged at every level of intellectual property sharing (Krishnamurthy, 2008). Open access has the potential to expand avenues for both publishing and accessing scholarly material. Open access creates exciting new opportunities and faces several formidable challenges. Open access has already changed the work of many academic librarians, depending on their institutional context, discipline foci, and personal affiliation. The American Research Libraries Association (ARL) Task Force defines "Open access is a cost effective way to disseminate and use information. It is an alternative to the traditional subscription-based publishing modal possible by new digital technologies and networked communication".

Moreover, open access models are beneficial in a number of ways (Antelman, 2004; Nicholas & Rowlands, 2005) as follows:

- Research is available at no cost, and with no access restrictions, for readers around the globe; scholars in economically disadvantaged areas are no longer barred access to the newest research.
- Because research published via avenues of open access is openly accessible online, it is more easily discoverable both by scholars and by search engines.
- For scholars in science and technology, where subject matter may be especially time- 6 sensitive, publication in open access journals occurs much more rapidly, and not necessarily with any impact on quality control.
- Since research is available globally without access restrictions, scholars benefit from having a significantly larger, more diverse audience.
- Increased exposure to research will also lead to more numerous citations. These benefits are judged to be more or less uncontroversial, unlike the issue of cost which will be discussed more fully in a later section on the economic impact of open access.

Most people who surf the web do not realize that the internet was developed by researchers in order to share research, and that commerce was only allowed online years later. But the original research purpose has not been forgotten, and committed researchers have steadily moved the free online sharing of research from the periphery to the mainstream of scholarly communication. Today most universities, libraries, and research funding agencies are providing OA, experimenting with it, or considering it. Its only opponents have been academic publishers. But today most academic publishers are also providing OA, experimenting with it, or considering it. Even the ones continuing to lobby against it are hedging their bets by taking steps toward adaptation. This is what deep institutional change looks like (Suber, 2012). It is therefore important to place the open access movement into perspective. No one can be against the ideal of open and free environments for basic research and learning. Open Access seemed, to many libraries, to be a way to mend a system that was at breaking point. What we have seen for a while is that research funders, governments and international organizations endorse open access as a creator of wealth for the society as a whole. This trend is growing and has the possibility to greatly accelerate the move towards full open access models (Eriksson, 2012).

## 6. OA is a Bane for India?

In India, various open access initiatives have been undertaken and are operational. Some more are in developmental stage. Even though the overall picture of open access publishing in India looks promising, it makes an unhappy situation for the journal as well as the publishing front. Minj et al. (n.d.) found that most of these online Indian open access journals do not comply with indexing standards of OA, i.e. the Open Access Initiative – Protocol for Metadata Harvesting (OAI-PMH) protocol and thus lie outside the OAI interoperability framework. The search and display interface of these journals revealed lack of support for field-based metadata search and display. A consequence of this is that in spite of their online presence, the articles in these journals tend to be less used, as they are not easily "discoverable" due to poor metadata and poor indexing (Sawant, 2009). Most of the Indian journals suffer from "low circulation – low visibility – low impact factor" syndrome. With many fewer paid regional or international subscriptions, these journals have limited visibility, restricted mainly to the members of the association. With this limited visibility, these journals are cited less frequently than their western counterparts. The low impact factor inhibits authors from submitting their quality work to the Indian journals. Thus, it is expected that with OA, Indian journals will be able to reach to a wider audience. At the same time, loss, if any, of paid non-member subscriptions is less likely to have a major effect on the economics of these journals (Sahu and Parmar, 2006).

Through two open ended questions we realized that, the major barriers to OA publishing in India, according to the participants were:

- Fear that revenue loss from print subscriptions if journal becomes OA.
- Lack of technological skills by people using the system like editors, reviewers etc.
- Lack of IT infrastructure or the funds to develop it.
- Other barriers include copyright issues, lack of management support, and need to continue print journals due to the low penetration of Internet among user base (Minj, Singhal, and Abraham, (n.d.).

The other factors include the following:

- High cost of ICTs infrastructures and connectivity and poor telecommunication infrastructure;
- Adequate Funding to make and promote the internet infrastructure in developing countries;
- Inadequate encouragement, lack of awareness and misconceptions of the existence and benefits; and
- Managing IPR (Copy right) through different publishing agreements.

## 7. Issues and Challenges in Open Access Movement

Numerous trends towards open access publication have emerged over the course of the past few years, confronting academic libraries with new challenges and presenting promising opportunities . The issues surrounding open access publishing are almost as numerous as the journals it concerns. The most challenging issue facing those who develop and manage library collections is how they will keep track of open access sources. Recent mass media attention on the scholarly publishing process has brought about both internal and external examination of key aspects in the process, they are:

- The extent and quality of peer review
- The technologies and cost associated with the capture and display of information

- The technologies and the costs associated with the various distribution channels
- The pricing policies and subsequent business exercised by the various types of scholarly publishers
- The ever-increasing expectations of the principal consumer with regard to the method of information delivery along with its cost, availability, and extend of access.
- The impact of new technologies and current publishing economics on other sector of the communication process, namely, authors, libraries and allied industries such aggregators and subscription agents.
- The technological challenges to ensuing the accuracy and accessibility of archives both present and future along with the potential impact of author archives and repository archives
- The overall sustainability of the scholarly communication process.

Shifting from the traditional model of scholarly communication to open access is a significant move, perhaps even a revolutionary one. There are numerous ways in which open access might impact an academic library, broken into the following categories as economic, technological, collection development & management, and the very roles that academic libraries play (Giarlo, 2005).

## 8. India's contribution to OA movement

It is heartening to note that the academic and scientific community in India have been contributing and promoting the concept of OA. Various Indian R&D organizations, leading scientific research institutions (such as Indian Institute of Science, IITs, ISI, institutes under the CSIR and Indian Council of Medical Research etc.) are now taking part in the open access movement by establishing institutional and digital repositories to provide worldwide access to their research literature. Several Indian publishers have already adopted the open-access philosophy for the electronic versions of their journals. Unlike some open access journals in other countries, in which authors pay to publish their papers, Indian open access journals use government grants and subscriptions to their print version to cover publishing costs (Arunachalam, 2004).

In the recent study by Bhardwaj (2015), identifies the major OA initiatives in India. It was found that 105 Indian repositories are listed in the Registry of Open Access Repositories (ROAR), constituting 3% of the total. The top 10 countries collectively produced 1,146,187 (63.4%) OA papers. Brazil has the maximum share of publications, 335,590 (18.6%), followed by United Kingdom 193,309 (10.7%), United States 115,078 (6.4%), and India 94,156 (5.2%). India publishes 590 OA journals and is ranked fourth in the world. The maximum OA journals are on medicine and majority of journals are in English 584 (97.8%). Indian researchers have been ranked eighth in research publications with 351 (3.6%) papers. Indian Institute of Science (IISc) is the leading institution in publishing research on OA. S. Arunachalam, from M.S. Swaminathan Research Foundation, Chennai is the most prolific Indian author publishing research on OA with 10 (2.8%) papers.

## 9. India's contribution in DOAJ by year wise

The data in Table 1 provides the complete representation of Indian contribution to DOAJ (Mondal, 2014; Ramesh Babu, 2015; Nashipudi, & Ravi, 2015). India has joined DOAJ in the year 2002. However, it started contribution from the year 2003 onwards. The maximum number of journals added in the year 2006 i.e. 206 (34.62) out of 590 journals. The Indian contribution from 2009 to 2013 constituting considerably increased. The data summarized in Table 1 show that India has made important contributions towards the growth of open access publishing. India's contribution with 590 open access journals as on 2016 is significant towards the growth of universe of knowledge. Indian contribution to the growth of universe of knowledge is optimum. This data would demonstrate the value of publishing in the OA journals and thereby giving their contributions for the growth of universe of knowledge.

**Table 1: Year wise Addition of Journals in DOAJ by India**

| S. No | Year | Journals | % |
|---|---|---|---|
| 1 | 2002 | 0 | 0 |
| 2 | 2003 | 12 | 2.01 |
| 3 | 2004 | 15 | 2.52 |
| 4 | 2005 | 12 | 2.01 |
| 5 | 2006 | 15 | 2.52 |
| 6 | 2007 | 15 | 2.52 |
| 7 | 2008 | 22 | 3.69 |
| 8 | 2009 | 38 | 6.55 |
| 9 | 2010 | 86 | 16.47 |
| 10 | 2011 | 59 | 11.25 |
| 11 | 2012 | 79 | 13.27 |
| 12 | 2013 | 206 | 34.62 |
| 13 | 2014 | 13 | 2.17 |
| 14 | 2015 | 16 | 0.32 |
| 15 | 2016 | 2 | 0.04 |
| | | 590 | 100 |

Source: DOAJ

## 10. Task Ahead to make OA a Boon for India

OA journals causing maximum access, visibility and impact to the research done in India. It could be concluded that open access helps to improve the accessibility of the journals. India is not only leading open access movement of the developing countries, but also making developed countries aware of qualitative scholarly literature originated from developing countries.

Finally, in order to tap the potential of OA, the following key *success factors* need to be considered:

- Implementation should produce economic value, i.e., the reduction of costs and savings of foreign currency and social value, i.e., a wider access to information and computer training.
- There needs to be enough trained people to support and use the OA solution. Training the users and developers needs to be given a high priority.
- Policy support for an OA strategy needs to expand to include all key players at the governmental and departmental levels, IT professionals, and computer users in general.
- Access to papers written by scientists within an institution in India is very scant, and, therefore, it is very important to locally start an effort to enhance access to scientific publications (Ramesh Babu, 2016).

## 11. Conclusion

Open access makes scholarly materials accessible to users at no cost. "A commitment to scholarly work carries with it a responsibility to circulate that work as widely as possible: this is the access principle. In the digital age, that responsibility includes exploring new publishing technologies and economic models to improve access to scholarly work. Wide circulation adds value to published work; it is a significant aspect of its claim to be knowledge. The right to know and the right to be known are inextricably mixed. Open Access can benefit both" (Willinsky, 2010). Open access provides a new paradigm for libraries and librarians. It provides unprecedented opportunities for librarians to deliver rich content in the quest to ensure that there is growth and development. It has the potential to deliver opportunities to loosen the noose of those communities that are strangled because of the lack of critical information. Open access is meant to return science to scientists themselves, but many leading researchers in the scientific community may either benefit from 'technology transfer' arrangements or see no conflict at all between commercialized research and the goals of open science. The open access movement itself offers neither an incentive nor a moral pressure for these researchers to tackle the more deeply rooted institutional obstacles to open and free scientific inquiry.

## References

Antelman, K. (2004). Do open-access articles have a greater research impact? *College & Research Libraries News*, 65(5), 372–382.

Arunachalam, Subbiah (2004). "India's March towards Open Access." *SciDev. Net*, 5 March 2004. Available at *http://www.scidev.net/quickguides/index.cfm?fuseaction=qguideReadItem&type=3&itemid=243&language=1&qguideid=4 (Accessed on 16th Feb. 2016)*

Association of Research Libraries. (2004). *Framing the issue: open access*. Available at: <http://www.arl.org/scomm/open_access/Framing_Issue_May04.pdf.

Berlin Declaration on Open Access (2003). Berlin Declaration on open access to knowledge in the sciences and humanities. Available at: http://oa.mpg.de/openaccess-berlin/berlindeclaration.html

Bethesda Statement (2003). Bethesda statement on open access publishing. Available at:

Bhardwaj, Raj Kumar (2015). India's Contribution to Open Access Movement, *Journal of Knowledge & Communication Management*, 5 (2) , 107–126.

Das, A.K and Ghosh, S.B. (2006).Open Access and Institutional Repositories-A developing Country Perspective: A case study of India. In, *World Library and Information Congress: 72nd IFLA General Conference and Council*, Seoul, Korea.

Eriksson, Jorgen (2012). From open access to openness. Available at *blog.inasp.info/open-access-openness. Accessed on 20 May 2016.*

Ghosh, S B and Das, A K (2006). Open access and institutional repositories – a developing country perspective: a case study of India . *World Library and Information Congress: 72ND IFLA General Conference and Council* 20-24 August 2006, Seoul, Korea. (Available at : http://www.ifla.org/IV/ifla72/index.htm ) accessed on 23 May 2016.

Giarlo, M.J. (2005). The impact of open access on academic libraries. Available at:

Gopalakrishnan, S and Ramesh Babu, B (2010).Open Source in Library and Information Science and IR Software. IN: *E-Publishing: UGC – SAP Symposium Proceedings.* Chennai: University of Madras. Pp. 72–98.

Harnad, S. (2008). Mandates and metrics: How open repositories enable universities to manage, measure and maximise their research assets. Available at:

http://lackoftalent.org/michael/papers/532.pdf (Accessed on 18 May 2016)

http://users.ecs.soton.ac.uk/harnad/Temp/openaccess.pdf (accessed on 14 May 2016)

http://www.earlham.edu/~peters/fos/bethesda.htm (accessed on 12 May 2016)

Krishnamurthy, M. (2008). Open access, open source and digital libraries: A current trend in university libraries around the world. *Program: electronic library and information systems*, 42 (1),48–55.

Minj, S., Singhal, M. and Abraham, T. (n.d.), Barriers to electronic publishing of scholarly journals from India: findings from the Scientific Journal Publishing in India (SJPI) project, Available at: http://eprints.rclis.org/archive/00014861/01/India_paper.pdf (accessed on 10 May 2016).

Mondal, D. (2014). India's contribution to DOAJ with special reference to Computer Science Discipline: A study. *International Journal of Information Dissemination and Technology*,4(1): 1–7.

Nashipudi, M. & Ravi, B. (2015). Contribution of India to Universe of Knowledge in DOAJ: A case study. *International Journal of Information Dissemination and Technology*, 5(3):171–175.

Nicholas, D. & Rowlands, I. (2005). Open access publishing: The evidence from the authors. *The Journal of Academic Librarianship*, 31(3),179–181.

Ramesh Babu , B (2016). Bridging the Digital Divide through Open Source and Open Access Movement: Issues, Challenges and Opportunities in India. IN: *UGC- SAP National Conference on Bridging the Digital Divide : Open Source and Open Access Movement, Proceedings* held on 11-12 March 2016 at the University of Mysore, edited by Mallinath Kumbar, NS Harinarayana and M Chandrasekara. Mysore: University of Mysore, DOS in LIS, pp. 9–17

Ramesh Babu, B (2015). Electronic Resource Management in the New Generation Academic Libraries: Issues and Challenges (Keynote address). IN: *Management of E-Resources in Degree College Libraries (NSMEDL 2015):* Seminar proceedings, edited by K .Somasekhara Rao, G. Prabhakar and L. Atchamamba. Vijayanagaram : Shri Durgaprasad Saraf College of Arts and Applied Sciences, pp. xviii–xxxviii.

Ramesh Babu, B and Nageswara Rao, P (2015). Issues and challenges of Open Access journals in India: Need for Digital Information Literacy. In: *Proceedings of the National Seminar on Role of LIS Professionals in developing Information Literacy Skills in the digital era*, edited by S. Thanskodi and R Jeyshankar. Karaikkudi: DLIS, Alagappa University, pp.126–132.

Sahu, D.K. (2006), "Open access publishing in the developing world: economics and impact", paper presented at *Asia Commons: Asian Conference on the Digital Commons*, 6-8 June 2006, Bangkok, Thailand, available at: http://openmed.nic.in/1598/01/Openaccess_Medknow.pdf (accessed 10 May 2016).

Sahu, D.K. and Parmar, R. (2006), "The position around the world: open access in India", in: Jacobs, N. (Ed.), *Open Access: Key Strategic, Technical and Economic Aspects*, Oxford :Chandos Publishing, pp. 26–32.

Sawant, Sarika (2009) "The current scenario of open access journal initiatives in India", *Collection Building*, 28 (4), 159–163

Suber, Peter (2003). Open Access to Science and Scholarship. Geneva: World Summit on the Information Society. Available at : http://www.earlham.edu/~peters/writing/wsis.htm (accessed on 2 April 2016)

Suber, Peter (2004). A very brief introduction to open access. Available at http://www.earlham.edu/~peters/hometoc.htm

Suber, Peter (2012). Open Access to research. Available at : https://www.opensocietyfoundations.org/voices/opening-access-research (Accessed on 20 May 2016)

Tise, Ellen R. ( 2010 )Open access: a new paradigm for libraries and a new role for librarians. Available at http://www.kb.se/dokument/Aktuellt/utbildning/ifla%20OA%202010/IFLA2010_EllenTise.pdf

Willinsky, J. (2003). Nine Flavors of Open Access Publishing. *Postgraduate Journal of Medicine*, *49*(3), 263--267.

Willinsky, J. (2010). The access principle: The case for open access to research and scholarship. Available at:http://mitpress.mit.edu/catalog/item/default.asp?tid=10611&ttype=2

# 4

# Scholarly Research Publications from India in Basic Medical Sciences, Engineering, Arts and Humanities as Reflected by the SCImago JCR Portal

**K. Veeranjaneuyulu[1], G. Rathinasabapathy[2] and S. Kopperundevi[3]**

[1] University Librarian, Prof.Jayasankar Telengana State Agricultural University, Hyderabad

[2] University Librarian, Tamil Nadu Veterinary & Animal Sciences University, Chennai

[3] Assistant Librarian, Veterinary College & Research Institute, Tirunelveli

**ABSTRACT**

*Scholarly research publications from India in the fields of Basic medical sciences, engineering, arts and humanities has been analysed using the Scimago Journal & Country Ranking Portal, a new size-independent indicator of scientific journal*

*prestige. Indian contributions in the following fields are significant as it is within the top 10 list as per data revealed by Scimago Journal & Country Ranking Portal viz., Multi-disciplinary (4th position with 21089 documents and h-index of 146); Chemical Engineering (5th position with 71796 documents and h-index of 219); Chemistry (5th position with 176897 documents and h-index of 253); Dentistry (6th position with 8105 documents and h-index of 53); Energy (6th position with 33146 documents and h-index of 149); Environmental Science (6th position with 71218 documents and an h-index of 190); Business, Management and Accounting (7th position with 18736 documents and h-index of 84); Engineering (8th position with 215550 documents and h-index of 210); Computer Science (9th position with 122005 documents and h-index of 162); Material Science (9th position with 150208 documents and h-index of 229); India's position in the following subject areas are not encouraging viz., Psychology (31), Arts and Humanities (23), Health Profession (21), Neuroscience (18), Immunology and Microbiology (12), Earth and Planetary Science (12), Medicine (12), Decision Science (11), Economics, Econometrics and Finance (11) and Mathematics (11).*

***Keywords:*** *Scientometrics, Journal Ranking, Country Ranking, Scholarly Publications, R&D Output, Citation, h-index, Basic medical sciences, Engineering, Arts, Humanities*

## 1. Introduction

World renowned scientist Albert Einstein once said "We owe a lot to the ancient Indians, teaching us how to count. Without which most modern scientific discoveries would have been impossible.". India, one of the oldest civilizations in the world, has a strong tradition of science and technology. Ancient India was a land of sages and seers as well as a land of scholars and scientists. Research has shown that from making the best steel in the world to teaching the world to count, India was actively contributing to the field of science and technology centuries long before modern laboratories were set up. Many theories and techniques discovered by the ancient Indians have created and strengthened the fundamentals of modern science and technology. While some of these groundbreaking contributions have been acknowledged, some are still unknown to most. India is playing a pivotal role in inventions and innovations in modern science, technology and medicine. In this background, the present study attempts to quantitatively analyse the contribution of India towards global scientific research publications in the disciplines of basic medical sciences, engineering, technology, multi-disciplinary, arts and humanities, etc., based on SCImago Journal & Country Rank Portal available in the public web.

### 1.1. SCImago Journal & Country Rank

SCImago Journal Rank (SJR indicator) is a measure of scientific influence of scholarly journals that accounts for both the number of citations received by a journal and the importance or prestige of the journals where such citations come from. The SJR indicator is a variant of the eigenvector centrality measure used in network theory. Such measures establish the importance of a node in a network based on the principle that connections to high-scoring nodes contribute more to the score of the node. The SJR indicator, which is inspired by the PageRank algorithm, has been developed to be used in extremely large and heterogeneous journal citation networks. It is a size-independent indicator and its values order journals by their "average prestige per article" and can be used for journal comparisons in science evaluation processes. The *SJR indicator* is a free journal metric which uses an algorithm similar to PageRank and provides an alternative to the impact factor (IF). Average citations per document in a 2-year period, abbreviated as "Cites per Doc. (2y)", is another index that measures the scientific impact of an average article published in the journal. It is computed using the same formula as the journal IF.

## 2. Objectives of the Study

The following are the objectives of the present study

- to quantitatively measure the publication output of countries in various disciplines *viz.*, basic medical sciences, engineering, technology, multi-disciplinary, arts and humanities using SCImago Journal & Country Rank portal
- to measure the quantity of India's contribution to the particular discipline
- to identify the top 10 countries in the discipline

## 3. Methodology

For this study, the SCImago Journal & Country Rank Web Portal has been used and the publication data pertaining to India has been downloaded. Then the data were analysed based on simple percentage analysis method and the results are interpreted.

## 4. Results and Discussion

### 4.1. Arts & Humanities

The study revealed that India is lagging behind in Arts and Humanities research publications as it is in 23rd position next to Finland. While China is in 10th position, India is in 23rd position with 7355 documents in which citable documents are 6775 and there are 62862 citations in which 12772 are self-citations and the citation per document is 8.55 and the h-index is 102. While there are more than 500 universities exclusively for Arts & Humanities in India, the share of our country in Arts & Humanities publication is not sufficient and the researchers and policy makers of higher education have to very seriously address this issue and take suitable remedial measure to get our deserving position at International level. The Top 10 countries based on documents published in Arts & Humanities are listed in Table - 1

**Table-1: Top 10 Countries with more publications in Arts & Humanities**

| Rank | Country | Documents | Citable documents | Citations | Self-citations | Citations per document | H index |
|---|---|---|---|---|---|---|---|
| 1 | United States | 379840 | 350199 | 10779998 | 5471158 | 28.38 | 930 |
| 2 | United Kingdom | 134514 | 123469 | 2637973 | 703798 | 19.61 | 476 |
| 3 | Germany | 54057 | 51219 | 1296037 | 285702 | 23.98 | 400 |
| 4 | France | 52716 | 50191 | 833297 | 157312 | 15.81 | 334 |
| 5 | Canada | 52544 | 49191 | 1175320 | 195305 | 22.37 | 363 |
| 6 | Australia | 39014 | 36675 | 648028 | 130775 | 16.61 | 269 |
| 7 | Spain | 33246 | 31868 | 328736 | 67496 | 9.89 | 200 |
| 8 | Italy | 31560 | 29847 | 549849 | 111712 | 17.42 | 261 |
| 9 | Netherlands | 29878 | 28147 | 729287 | 119034 | 24.41 | 299 |
| 10 | China | 20591 | 19902 | 221077 | 80215 | 10.74 | 190 |

## 4.2. Business, Management & Accounting

In the field of Business, Management and Accounting, India has contributed 18736 documents in which 18383 are citable and earned 71279 citations. In this, 20261 are self-citations and the citations per document are 11.2 and the h-index is 234 and is in 7th position. The details are depicted in Figure-1.

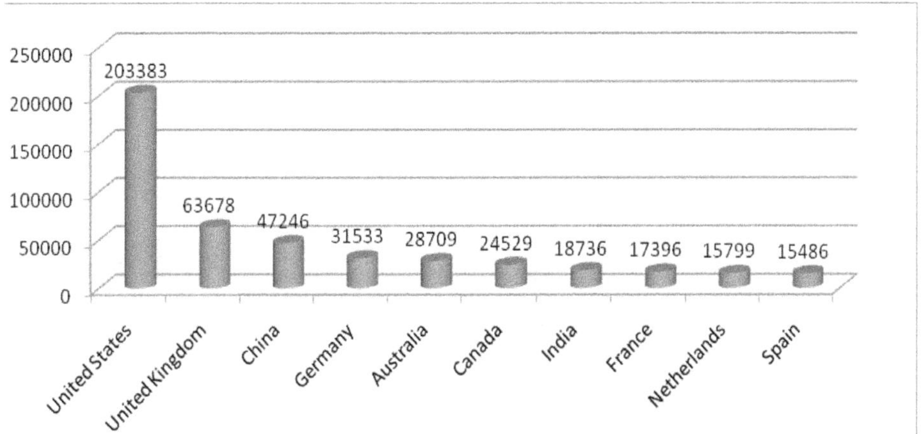


**Fig. 1 Top 10 Countries in Business, Management & Accounting Publications**

## 4.3. Chemical Engineering

India is in 5th place as far as the number of documents is published in Chemical Engineering with 71796 documents. In this, 70929 documents are citable and it has earned 865654 citations so far in which 280599 are self-citations and the citations per document is 12.06 with an h-index of 219. In this field also USA is in the top followed by China in second position. The details are furnished in Table-2.

**Table-2: Top 10 Countries with more publications in Chemical Engineering**

| Rank | Country | Documents | Citable documents | Citations | Self-citations | Citations per document | H index |
|---|---|---|---|---|---|---|---|
| 1 | United States | 286373 | 277018 | 7473750 | 2692676 | 26.1 | 600 |
| 2 | China | 260153 | 258717 | 2672314 | 1505137 | 10.27 | 313 |
| 3 | Japan | 98449 | 96863 | 2039504 | 564929 | 20.72 | 323 |
| 4 | Germany | 92522 | 90611 | 1995543 | 485436 | 21.57 | 356 |
| 5 | India | 71796 | 70929 | 865654 | 280599 | 12.06 | 219 |
| 6 | United Kingdom | 70660 | 68631 | 1626933 | 324550 | 23.02 | 327 |
| 7 | France | 61755 | 60716 | 1326575 | 296543 | 21.48 | 281 |
| 8 | South Korea | 57770 | 57275 | 854125 | 189860 | 14.78 | 242 |
| 9 | Canada | 46495 | 45632 | 903686 | 169494 | 19.44 | 257 |
| 10 | Spain | 44201 | 43573 | 904416 | 213260 | 20.46 | 227 |

## 4.4. Chemistry

Like Chemical Engineering, India is in 5th position in Chemical Engineering discipline as well. India's contribution to this field is 176897 documents in which 175185 are citable and these documents were cited 2048623 times in which 778963 are self-citations and the citations per document is 11.58 and the h-index 253. It has been noted that India is in 5th place in Chemistry as well as in Chemical Engineering while USA and China are in the first and second position respectively. The details are depicted in Figure-2.

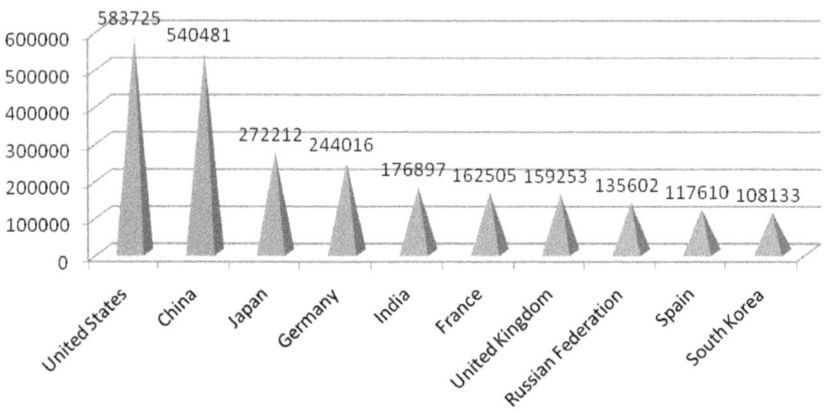


**Fig.2 Top 10 Countries with more documents in Chemistry**

## 4.5. Computer Science

It is a known fact that India is known for its contribution in the field of Computer Science and Information Technology. As far as the publications in computer science, India is in 9th position with 122005 documents in which 120389 are citable. There are 457742 citations for Indian publications in which 126373 are self-citations and the citations per document is 3.75 and the h-index is 162. As seen in other fields, USA and China are in the top two places. The details are depicted in Figure-3

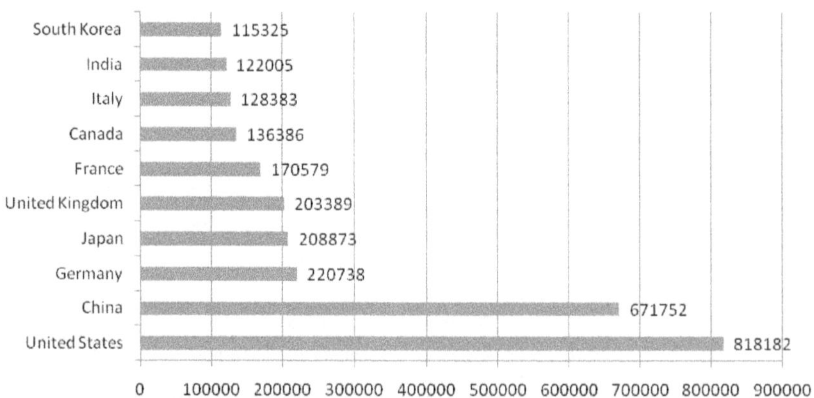


**Fig. 3 Top 10 Countries with more documents in Computer Science**

## 4.6. Decision Science

Decision science is a collaborative approach involving mathematical formulae, business tactics, technological applications and behavioural sciences to help senior management make data driven decisions. The performance of India in decision science publications is not up to the mark as it is in 11th position with 8733 documents in which 8558 documents are citable. The total citations for Indian publications is 146280 in which 23302 are self-citations and the citations per document is 7.45 and the h-index in 88. The details are depicted in Figure-4.

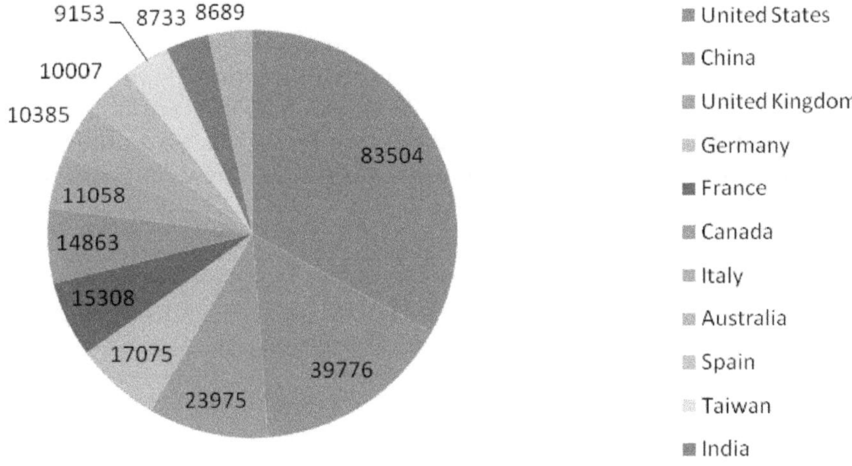


**Fig. 4 Top 10 Countries with more Publications in Decision Science**

## 4.7. Dentistry

India has contributed 8105 documents in the field of dentistry in which 7415 are citable documents and the total citations received are 33357. While the self-citations is 9809, the citations per document is 4.12 and the h-index is 53. With this contribution, India is in 6th position followed by Germany while the first and second positions are taken by USA and Brazil. The details are furnished in Table -3.

**Table – 3: Top 10 Countries with more documents in Dentistry**

| Rank | Country | Documents | Citable documents | Citations | Self-citations | Citations per doc. | H index |
|---|---|---|---|---|---|---|---|
| 1 | United States | 41707 | 38201 | 679318 | 235369 | 16.29 | 191 |
| 2 | Brazil | 16333 | 16054 | 164314 | 59925 | 10.06 | 100 |
| 3 | United Kingdom | 16141 | 14477 | 237724 | 49672 | 14.73 | 130 |
| 4 | Japan | 16026 | 15810 | 193746 | 48100 | 12.09 | 113 |
| 5 | Germany | 9209 | 8862 | 149725 | 31570 | 16.26 | 109 |
| 6 | India | 8105 | 7415 | 33357 | 9809 | 4.12 | 53 |

(Continued)

**Table – 3 (Continued)**

| Rank | Country | Documents | Citable documents | Citations | Self-citations | Citations per doc. | H index |
|---|---|---|---|---|---|---|---|
| 7 | Italy | 7340 | 7009 | 101685 | 21158 | 13.85 | 103 |
| 8 | Turkey | 6325 | 6183 | 64933 | 12146 | 10.27 | 71 |
| 9 | China | 5069 | 4944 | 47685 | 11864 | 9.41 | 66 |
| 10 | Sweden | 4672 | 4581 | 112082 | 17627 | 23.99 | 115 |

## 4.8. Earth and Planetary Science

The contribution of India in this discipline is not up to the mark as it is in 12th position followed by Spain. India has 46378 documents in which 45115 are citable and the total citations are 354099 in which 155076 are self-citations. The citations per documents are 7.64 and the h-index is 132. Like other disciplines, the USA, China and UK are in the top 3 positions. The details are depicted in Figure-5.

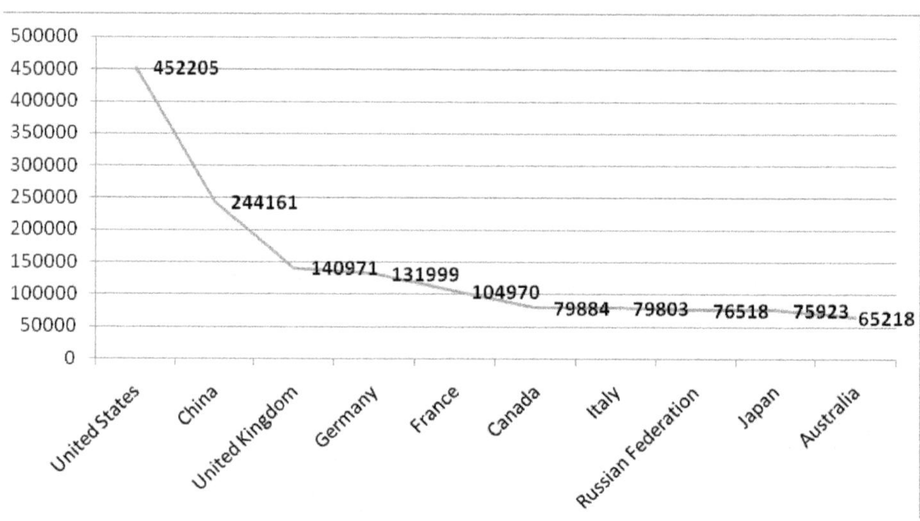


**Fig. 5: Top 10 Countries with more documents in Earth and Planetary Science**

## 4.9. Economics, Econometrics and Finance

The study revealed that India is in 11th place as far as the research publications in this discipline are concerned. The total number of documents published form India is 9556 in which 8703 are citable and received 40784 citations in which 10590 are self-citations. The citations per document are 4.27 and the h-index is 73. The top three countries are USA, UK and Germany. The details are furnished in Table-4.

**Table-4 Top 10 Countries with more documents in Economics, Econometrics & Finance**

| Rank | Country | Documents | Citable documents | Citations | Self-citations | Citations per document | H index |
|---|---|---|---|---|---|---|---|
| 1 | United States | 146539 | 139259 | 2869830 | 1273949 | 19.58 | 430 |
| 2 | United Kingdom | 46838 | 44470 | 689879 | 175073 | 14.73 | 224 |
| 3 | Germany | 26696 | 25604 | 251135 | 60031 | 9.41 | 137 |
| 4 | France | 20647 | 19860 | 168437 | 30009 | 8.16 | 127 |
| 5 | Canada | 19347 | 18531 | 261080 | 38422 | 13.49 | 162 |
| 6 | Australia | 18761 | 18104 | 184473 | 42718 | 9.83 | 127 |
| 7 | China | 15464 | 15011 | 100809 | 41690 | 6.52 | 99 |
| 8 | Italy | 14612 | 14127 | 138335 | 29966 | 9.47 | 117 |
| 9 | Spain | 14483 | 14076 | 129880 | 27534 | 8.97 | 111 |
| 10 | Netherlands | 13204 | 12673 | 208423 | 31890 | 15.78 | 148 |
| 11 | India | 9556 | 8703 | 40784 | 10590 | 4.27 | 73 |

## 4.10. Energy

Indian contribution in the discipline of Energy is 215550 documents and it is in 6th place followed by South Korea while the top three places are occupied by China, USA and Japan. The citable documents of India are 32585 with 279436 citations which includes 84130 self-citations. The citations per documents are 8.43 and the h-index is 149. In the field of Energy, India is within the top 10 countries which produced more documents. The details of publications in Energy are depicted in Figure-6.

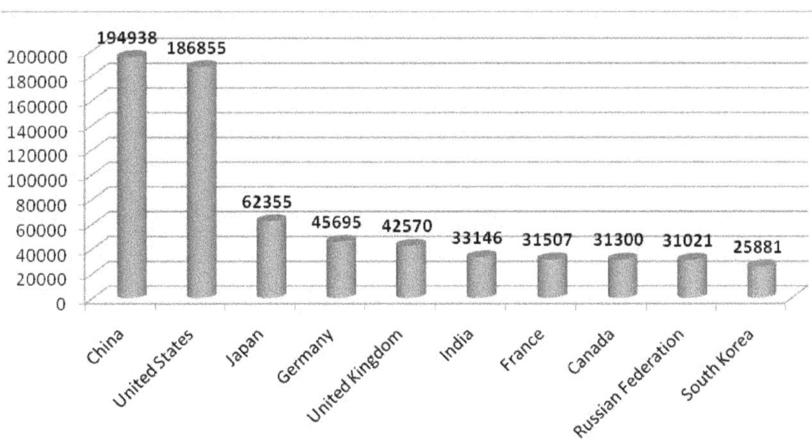


**Fig. 6: Top 10 Countries with more documents in Energy**

## 4.11. Engineering

The SCImagor Journal & Country Rank portal data revealed that India is in 8th position as far as publications in Engineering are concerned. With 215550 documents in which 213073 are citable, India has received 1202720 citations in which 388154 are self-citations and the citations per document is 5.58 and the h-index is 210. The details of Engineering publications of top 10 countries are listed in Table-5.

**Table-5: Top 10 Countries with more documents Engineering**

| Rank | Country | Documents | Citable documents | Citations | Self-citations | Citations per document | H index |
|---|---|---|---|---|---|---|---|
| 1 | China | 1520514 | 1513451 | 5300669 | 3441224 | 3.49 | 334 |
| 2 | United States | 1503616 | 1471087 | 15099037 | 5710411 | 10.04 | 708 |
| 3 | Japan | 517393 | 511476 | 3153319 | 1050082 | 6.09 | 319 |
| 4 | Germany | 366061 | 359882 | 2872260 | 719190 | 7.85 | 357 |
| 5 | United Kingdom | 334595 | 326430 | 3251695 | 715559 | 9.72 | 375 |
| 6 | France | 257584 | 254247 | 2186642 | 566652 | 8.49 | 301 |
| 7 | South Korea | 242461 | 239592 | 1732964 | 398314 | 7.15 | 255 |
| 8 | India | 215550 | 213073 | 1202720 | 388154 | 5.58 | 210 |
| 9 | Canada | 210198 | 206390 | 1957416 | 373777 | 9.31 | 307 |
| 10 | Italy | 200629 | 197108 | 1770205 | 479375 | 8.82 | 268 |

## 4.12. Environmental Science

The study revealed that with a total number of 71218 documents, India is in 6th place in Environmental Science publications following Canada while the USA, China and UK are the top three countries. Out of 71218 documents, 69517 are citable documents and they were cited by 637503 in which 223794 are self-citations. The citations per documents are 8.95 and the h-index is 190. The details of documents published by top 10 countries in Environmental Sciences are listed in Table-6.

**Table-6: Top 10 Countries with more documents in Environmental Science**

| Rank | Country | Documents | Citable documents | Citations | Self-citations | Citations per document | H index |
|---|---|---|---|---|---|---|---|
| 1 | United States | 433231 | 417603 | 8164911 | 3916606 | 18.85 | 487 |
| 2 | China | 185217 | 182794 | 1618324 | 930727 | 8.74 | 215 |
| 3 | United Kingdom | 121873 | 117137 | 2538532 | 683448 | 20.83 | 332 |
| 4 | Germany | 97290 | 94084 | 1607020 | 399956 | 16.52 | 287 |
| 5 | Canada | 82331 | 79795 | 1549382 | 398587 | 18.82 | 271 |
| 6 | India | 71218 | 69517 | 637503 | 223794 | 8.95 | 190 |
| 7 | Japan | 68095 | 66730 | 876287 | 220043 | 12.87 | 196 |
| 8 | France | 66642 | 65119 | 1244511 | 305037 | 18.67 | 257 |
| 9 | Australia | 65779 | 63561 | 1175082 | 362102 | 17.86 | 246 |
| 10 | Spain | 56817 | 55732 | 1023631 | 294917 | 18.02 | 214 |

## 4.13. Health Professions

It has been observed that India's position in publications dealing with health professions is 12th following Australia while the top three places are occupied by the USA, UK and Canada. The study revealed that India has 32817 documents in which 31624 are citable and the total citations are 380133 in which 129007 are self-citations. The citations per document is 11.58 and the h-index is 146. The details are depicted in Figure-6.

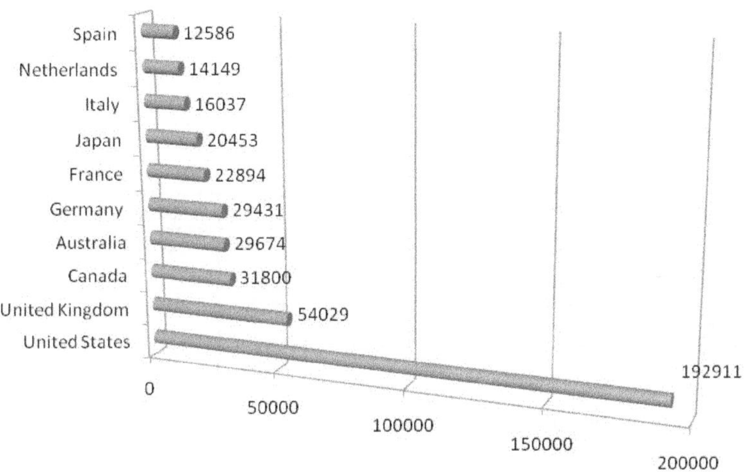


**Fig-6: Top 10 Countries with more documents in Health Professions**

## 4.14. Immunology and Microbiology

The total number of documents published by India in this field is 32817 through which it has secured 12th place following Australia while the first three places are secured by the USA, UK and Germany. While the citable documents of India are 31624, the total citations are 380133 in which 129007 are self-citations and the citations per document are 11.58 with h-index of 146. The particulars of documents published by the top 10 countries in Immunology and Microbiology are depicted in Figure – 7.

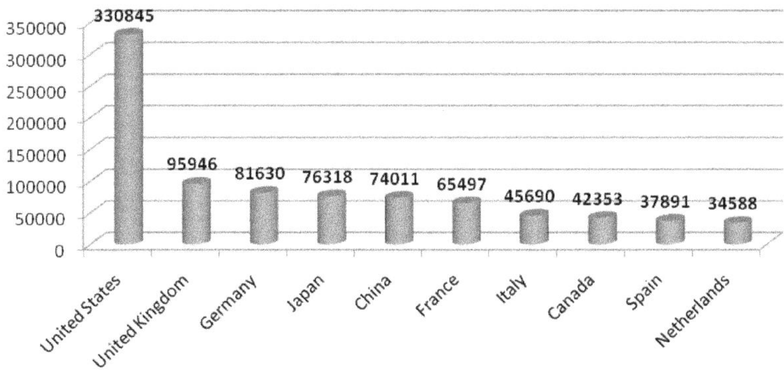


**Fig. 7: Top 10 Countries with more documents in Immunology and Microbiology**

## 4.15. Material Science

In material science research publication output, India is in 9th place with 150208 documents in which 148536 are citable and the total citations are 1522241 in which 534232 are self-citations. The citations per document are 10.13 while the h-index is 229. The details are listed in Table – 7.

**Table-7: Top 10 Countries with more documents in Material Science**

| Rank | Country | Documents | Citable documents | Citations | Self-citations | Citations per document | H index |
|---|---|---|---|---|---|---|---|
| 1 | United States | 716970 | 706193 | 11920604 | 4391609 | 16.63 | 643 |
| 2 | China | 705309 | 701235 | 5552920 | 3215349 | 7.87 | 342 |
| 3 | Japan | 357147 | 353735 | 4168936 | 1350113 | 11.67 | 353 |
| 4 | Germany | 283964 | 280591 | 3984115 | 1093917 | 14.03 | 382 |
| 5 | France | 187957 | 186256 | 2718179 | 698306 | 14.46 | 311 |
| 6 | United Kingdom | 177385 | 174377 | 2864905 | 597171 | 16.15 | 350 |
| 7 | South Korea | 160523 | 159114 | 1786327 | 434381 | 11.13 | 263 |
| 8 | Russian Federation | 159975 | 158777 | 862008 | 305234 | 5.39 | 191 |
| 9 | India | 150208 | 148536 | 1522241 | 534232 | 10.13 | 229 |
| 10 | Italy | 104449 | 103185 | 1370287 | 352079 | 13.12 | 238 |

## 4.16. Mathematics

India is known for its contribution towards mathematics since time immemorial. With its contribution of '0' to the world, India is always respected by the rest of the world and it has produced many world class mathematicians like Ramanujan. However, the contribution of India towards mathematics in recent years is not up to the mark as it is standing at 11th place after Spain while the USA, China and Germany are in the top three places. The details of top 10 contributors in mathematics are listed in Table- 8.

**Table – 8: Top 10 Countries with more documents in Mathematics**

| Rank | Country | Documents | Citable documents | Citations | Self-citations | Citations per document | H index |
|---|---|---|---|---|---|---|---|
| 1 | United States | 539376 | 529133 | 6070408 | 2447963 | 11.25 | 558 |
| 2 | China | 365826 | 362413 | 1594106 | 1042350 | 4.36 | 212 |
| 3 | Germany | 175077 | 171737 | 1480654 | 421145 | 8.46 | 257 |
| 4 | France | 151758 | 149144 | 1284435 | 410522 | 8.46 | 233 |
| 5 | United Kingdom | 143578 | 140037 | 1510710 | 372770 | 10.52 | 290 |
| 6 | Japan | 117898 | 116211 | 682176 | 218923 | 5.79 | 176 |
| 7 | Italy | 108139 | 105843 | 827778 | 267804 | 7.65 | 189 |
| 8 | Canada | 89747 | 87939 | 808876 | 163043 | 9.01 | 214 |
| 9 | Russian Federation | 83162 | 82026 | 299688 | 116731 | 3.6 | 119 |
| 10 | Spain | 81710 | 80037 | 606941 | 187744 | 7.43 | 175 |

## 4.17. Medicine

The study revealed that India is in 12th place following the Netherlands as far as the documents published in Medicine are concerned. It has published 232767 documents in which 199319 are citable and earned 1716085 citations so far in which 506943 are self-citations. The USA, UK and Germany are in the top 3 positions. The citations per document are 7.37 with h-index of 242. The details are depicted in Figure-8.

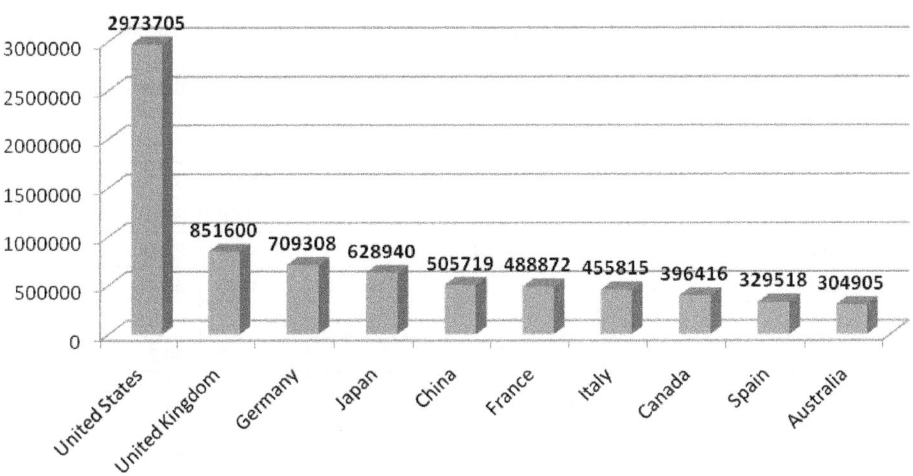


**Fig. 8: Top 10 countries with more documents in Medicine**

## 4.18. Multi-disciplinary

India's publications in multi-disciplinary subjects are laudable as it is in 4th place with 21089 documents in which 17688 are citable. The total citations are 171646 in which 66106 are self-citations and the citations per document are 8.14 with an h-index of 146. The USA, China and UK are the top three countries in this discipline. The details are depicted in Figure-9.

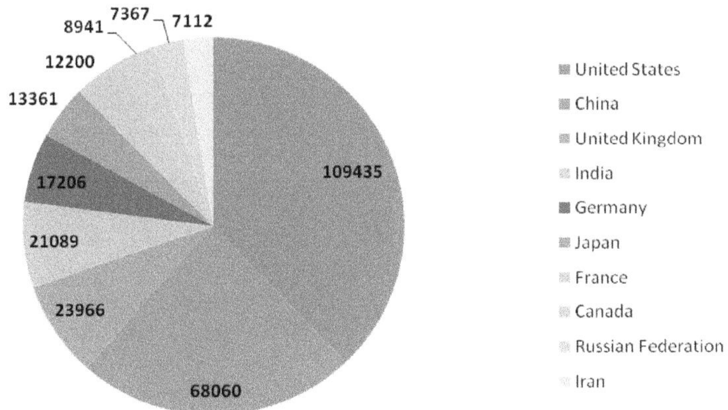


**Fig. 9: Top 10 Countries with more documents in Multidisciplinary**

## 4.19. Neuroscience

India's contribution towards Neuroscience research publications is not up to the mark as it is in 18th position following Belgium while the top three places are occupied by the USA, UK and Germany. India with 12169 documents to its credit in which 10182 are citable documents, earned 128233 citations so far in which 30588 are self-citations. The citations per document are 10.54 and the h-index is 101. The details of neuroscience publications by the top 10 countries are listed in Table-9.

**Table-9: Top 10 Countries with more documents in Neuroscience**

| Rank | Country | Documents | Citable documents | Citations | Self-citations | Citations per document | H index |
|---|---|---|---|---|---|---|---|
| 1 | United States | 353766 | 334403 | 12319512 | 6338455 | 34.82 | 650 |
| 2 | United Kingdom | 91518 | 84768 | 3074416 | 698259 | 33.59 | 420 |
| 3 | Germany | 89016 | 84060 | 2676952 | 652839 | 30.07 | 361 |
| 4 | Japan | 68192 | 65656 | 1561979 | 356708 | 22.91 | 280 |
| 5 | Canada | 56185 | 53249 | 1697336 | 314685 | 30.21 | 339 |
| 6 | Italy | 48204 | 44991 | 1196044 | 248192 | 24.81 | 280 |
| 7 | France | 47002 | 44631 | 1419301 | 255260 | 30.2 | 315 |
| 8 | China | 44719 | 43164 | 465490 | 166617 | 10.41 | 156 |
| 9 | Australia | 31775 | 29935 | 739520 | 142028 | 23.27 | 231 |
| 10 | Netherlands | 30000 | 28292 | 916147 | 149508 | 30.54 | 262 |

## 4.20. Nursing

Similar to Neuroscience, India's position in nursing publications is also not up to the mark as it is in 21st position following Belgium while the top 3 positions are secured by the USA, UK and Australia. India with 4126 documents to its credit in which 3933 are citable documents, has secured 50186 citations in which 15905 is self-citations. The citations per document are 11.9 while the h-index is 82. The details of top 10 countries with more documents in nursing are depicted in Figure-10.

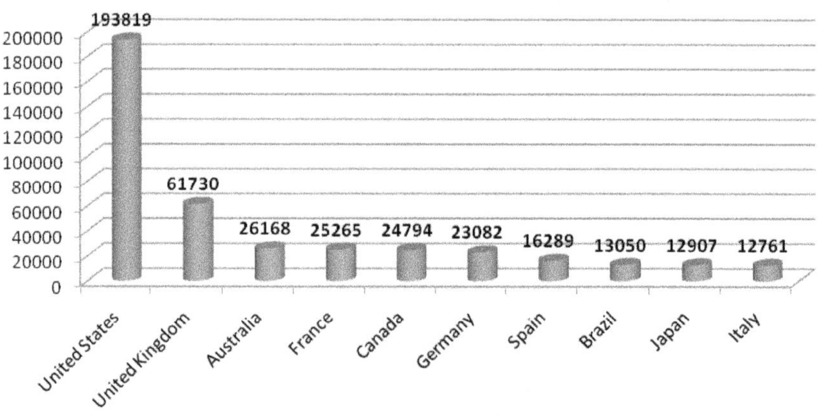


**Fig. 10: Top 10 Countries with more documents in Nursing**

## 4.21. Pharmacology, Toxicology and Pharmaceutics

The study revealed that India is really in a very good position in the field of Pharmacology, Toxicology and Pharmaceutics as it is in 3rd position following the USA and China with 99301 documents in which 96653 are citable documents and the total citations of 849882 in which 365880 are self-citations. The citations per document is 8.56 and the h-index is 181. The details are furnished in Table-10.

**Table-10: Top 10 Countries with more documents in Pharmacology, Toxicology and Pharmaceutics**

| Rank | Country | Documents | Citable documents | Citations | Self-citations | Citations per document | H index |
|---|---|---|---|---|---|---|---|
| 1 | United States | 312577 | 296644 | 7364021 | 3363136 | 23.56 | 486 |
| 2 | China | 125336 | 123831 | 970330 | 459726 | 7.74 | 165 |
| 3 | India | 99301 | 96653 | 849882 | 365880 | 8.56 | 181 |
| 4 | Japan | 97987 | 96326 | 1592754 | 424310 | 16.25 | 225 |
| 5 | United Kingdom | 83393 | 76172 | 1904301 | 387544 | 22.84 | 319 |
| 6 | Germany | 69750 | 66597 | 1412372 | 314753 | 20.25 | 270 |
| 7 | Italy | 50631 | 48855 | 999599 | 231242 | 19.74 | 215 |
| 8 | France | 48828 | 47217 | 1051245 | 199497 | 21.53 | 249 |
| 9 | Spain | 35845 | 33259 | 587681 | 134941 | 16.4 | 175 |
| 10 | Canada | 35492 | 33981 | 816871 | 140785 | 23.02 | 224 |

## 4.22. Physics and Astronomy

Indian contribution to Physics and Astronomy is considerably good as it is standing in 9th position following Italy with 169208 documents in which 167318 are citable documents. The total citations for these publications are 1627576 in which 564926 are self-citations and the citations per document is 9.62 and the h-index is 276. The details of top 10 countries withmore documents in physics and astronomy are furnished in Table-11.

**Table 11: Top 10 Countries with more documents in Physics & Astronomy**

| Rank | Country | Documents | Citable documents | Citations | Self-citations | Citations per document | H index |
|---|---|---|---|---|---|---|---|
| 1 | United States | 1072927 | 1058312 | 20629889 | 8975942 | 19.23 | 791 |
| 2 | China | 708744 | 704502 | 4954827 | 2835169 | 6.99 | 335 |
| 3 | Japan | 458257 | 453996 | 5528292 | 1909959 | 12.06 | 430 |
| 4 | Germany | 446100 | 441060 | 7544235 | 2383993 | 16.91 | 502 |
| 5 | France | 309552 | 306265 | 4915267 | 1371938 | 15.88 | 420 |
| 6 | United Kingdom | 290349 | 285854 | 5286895 | 1351162 | 18.21 | 466 |
| 7 | Russian Federation | 282416 | 279983 | 2307643 | 771100 | 8.17 | 314 |
| 8 | Italy | 212754 | 210216 | 3169433 | 932427 | 14.9 | 364 |
| 9 | India | 169208 | 167318 | 1627576 | 564926 | 9.62 | 246 |
| 10 | South Korea | 165904 | 164259 | 1784548 | 443992 | 10.76 | 276 |

## 4.23. Psychology

India is lagging behind in the field of Psychology as it is in 31st position following Ireland while the USA, UK and Canada are the top three countries. India's contribution in this field is a mere 3332 documents in which 3078 are citable. The total citations for these documents are 27860 in which 5582 are self-citations. The citations per document is 8.36 and the h-index is only 64. Though India has renowned institutes like NIMHANS, its research literature production is not up to the mark and the researchers and policy makers should serious look into this for suitable remedial measure. The details are depicted in Figure-11.

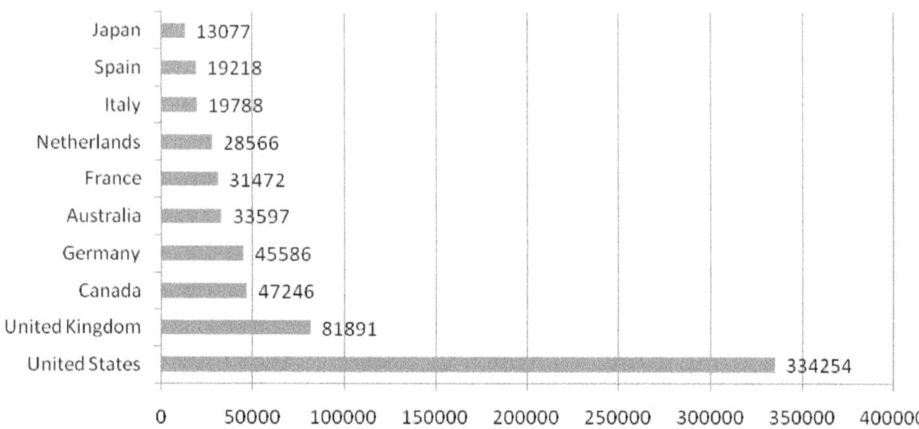


**Fig. 11: Top 10 Countries with more documents in Psychology**

# 5. Findings of the Study

- In Arts and Humanities, it is in 23rd position with 7355 documents and the citation per document is 8.55 and with h-index of 102.
- In Business, Management and Accounting, it is in 7th position with 18736 documents, citation per document is 3.8 and h-index is 84.
- India's position in Chemical Engineering is 5th with 71796 docuemnts, citations per document 12.06 with h-index of 219.
- In Chemistry, India again in 5th position with 176897 documents, citations per document is 11.58 with h-index of 253.
- India is playing a vital role in computer science however its position in scholarly research publications is 9th with 122005 documents, citation per documents 3.75 with h-index of 162.
- In Decision Science, India is in 11th place with 8733 documents and h-index of 88. With 8105 documents, 4.12 citations per document and h-index of 53
- India is in 6th position in Dentistry. Its position in Earth and Planetary Science is 12th with 46378 documents, 7.64 citations per document and 132 h-index.

- In Economics, Econometrics and Finance, India is in 11th position with 9556 documents and its h-index is 73.
- India with 33146 documents and h-index of 149 secured 6th position in Energy related publications while in 8th place in Engineering with 215550 documents with h-index of 210.
- Its position in Environmental Science is 6th with 71218 documents and an h-index of 190.
- Indian contribution to health profession publications is not encouraging as it is in 21st position with 4915 documents with h-index of 59.
- In Immunology and Microbiology, it has contributed 32817 documents with h-index of 146 and secured 12th position.
- With 150208 documents and h-index of 229, India is in 9th position in Material Science.
- Its position in Mathematics publications is also not up to the mark as it has not secured the top 10 position and it is in 11th position with 64615 documents and its h-index is 127.
- India's position in Medicine is also not encouraging as it is in 12th position with 232767 documents and its h-index is 242.
- In multi-disciplinary publications, India is in a good position viz., 4th place with 21089 documents and h-index of 146 following USA, China and U.K.
- In Neuroscience, it is in 18th position with 12169 documents and h-index is 101 and this needs to be look into by the concerned.
- India is lagging behind in the field of Psychology publications as it is in 31st place with hardly 3332 documents and h-index of 64 where many small countries are doing well in this area.
- In Social Science, India is in 11th position with 31185 documents and the h-index is 92 and India can do better in this area also.

## 6. Conclusion

As reflected by Scimago Journal and Country Ranking portal, Indian contribution to the following fields are significant as it is within the top 10 list *viz.*, Multi-disciplinary, Chemical Engineering, Chemistry, Dentistry, Energy, Environmental Science, Business, Management and Accounting, Engineering, Biochemistry, Genetics and Molecular Biology, Computer Science, Material Science and Agricultural Sciences. But, it is not encouraging in the following disciplines *viz.*, Arts and Humanities (23rd position with 7355 documents), Decision Science (11th place with 8733 documents), Economics, Econometrics and Finance (11th position with 9556 documents), Health profession (21st position with 4915 documents), Immunology and Microbiology (12th position with 32817 documents), Mathematics (11th position with 64615 documents), Medicine (12th position with 232767 documents), Neuroscience (18th position with 12169 documents), Psychology (31st place with hardly 3332 documents) and Social Science (11th position with 31185 documents).

This study clearly revealed that the concerned agencies responsible for education and research should take suitable measures to augment research activities and quality publications in the fields where India is not doing well and it is necessary to take India at least in the top 10 list of all these disciplines.

## References

Bornmann, Lutz, Felix de Moya-Anegón, and Loet Leydesdorff. "The new excellence indicator in the World Report of the SCImago Institutions Rankings 2011." *arXiv preprint arXiv:1110.2305* (2011).

Falagas, Matthew E., et al. "Comparison of SCImago journal rank indicator with journal impact factor." *The FASEB journal* 22.8 (2008): 2623–2628.

Jeremic, Veljko, et al. "Excellence with leadership: the crown indicator of Scimago Institutions Rankings Iber Report." *El profesional de la información* 22.5 (2013): 474–480.

Spiroski, Mirko. "Country rank, journal rank and H-index in the field of Medicine in the Republic of Macedonia (1996–2008) using data from Scimago." *Macedonian Journal of Medical Sciences* 3.2 (2010): 99–108.

Vicente P. Guerrero-Botea and Félix Moya-Anegónb (2012). A further step forward in measuring journals' scientific prestige: The SJR2 indicator Journal of Informetrics 6 (2012) 674–688 Available at http://www.scimagojr.com/files/SJR2.pdf Accessed on 15.10.2016.

Zacca-González, Grisel, et al. "Bibliometric analysis of regional Latin America's scientific output in public health through SCImago journal & country rank." *BMC public health* 14.1 (2014): 632.

# 5

# Digital Initiatives for Agricultural Research & Education under ICAR in India

**Amrender Kumar[#], Rakhee Sharma, Ashish Sharma, Kamal Batra, Sanjiv Kapur**

Agricultural Knowledge Management Unit,
ICAR-Indian Agricultural Research Institute (ICAR-IARI), New Delhi

**ABSTRACT**

*The usage of information, communication and dissemination system (ICDC) has a vast potential to transform the agricultural knowledge available in the various National Agricultural Research and Education System (NARES) into information for an efficient management. For achieving this, ICAR has initiated digital initiatives to capture and manage knowledge in various field of agriculture using ICT technology. Success of ICT enabled platform, largely depend on potential for acceptance by researchers, faculty, students and other stakeholders. These developed digital platforms improves the access to appropriate knowledge at opprtunate time which leads to smart solutions, which bridges the gap between agricultural researchers, extension agents and farmers hereby enhancing the agricultural output research in terms of quality and quantity.*

***Keywords:*** *Digital repository, Information and Communication Technology, Open access, Krishikosh and Institutional repository*

# 1. Introduction

Timely access to information is becoming more and more crucial for survival in every sphere of life and agriculture sector is no exception. In the present competitive word, moving towards what we perceive as knowledge society, the access to right information at anytime, anywhere, about anything has gained high significance. This off course does not mean that the earlier societies were not aware of importance of information or were not knowledgeable. The information played very important role even in ancient time when hunters & gather of the sub-continent evolved into agri-pastoral society, domesticated plants, animals and learned farming using draft animals, inventing tillage, seeding, intercultural operations, harvesting, and primary processing and prospered as interregional/international traders. They were knowledgeable enough to evolve into present day society. The crucial difference now is the speed with which you can access information, the magnitude of available information and removal of geographical boundaries to access information. The developments in computer technology itself revolutionized the world and the sudden growth in telecommunication methodologies provided the necessary synergy to create a catastrophic change breaking every boundary and connecting the planet into one giant network of information and knowledge.

Today, the information has become absolutely important input in agriculture along with seed, fertilizers, pesticides, land, water and environment. The contribution of public extension system in disseminating information and attaining self-reliance in food production is very well recognized. But in this changing time, traditional public extension system is not sufficient to address multi faceted problems faced by farmers. The existing public extension system is also constrained by limited resources, wide ratio between farmers and extension workers and also by added responsibility of handling emerging issues like marketing extension, agri-business, quality conscious consumers and WTO. The need of the hour is to evolve a comprehensive agriculture extension system shared by all the stakeholders. These stakeholders could be farmers co-operatives, progressive farmers, agricultural consultants, consultancy firms, farmers organizations, unemployed agricultural graduates, non-governmental organizations, Krishi Vigyan Kendras (KVKs –Farm Science Centers), agri-business companies, input dealers, newspapers, agricultural magazines, private television channels, private sector banks, market information systems, weather forecasting agencies, agro-advisory services etc.

The convergence of computers and communication technologies has open up vast arena of Internet and Intranet. One cannot ignore the silent revolution taking place in the communication systems in Rural India, thus, paving way for "Cyber Extension" initiatives. Concept of Village Information Kiosks is fast spreading to blocks/mandals and villages empowering Indian farmers to digitized access of vital information available through the Internet. There are dozens of cyber-experiments going on in rural India, which have unequivocally demonstrated the power of Internet and Information Technology. The overwhelming response and eagerness of farmers to use such systems is now paving the way for replicating cyber extension initiatives in large numbers. In such a scenario, the demand for authentic and credible digital information sources has risen in agriculture sector especially in research, education or extension. End to end value chain development requires quick access to diverse type of information.

In the present era of knowledge revolution the organization, capturing, preserving and reusing of knowledge has become absolutely essential for any organization to keep itself competitive and efficient. Digital repository with open access policy may cater the needs of National Agricultural Research and Education System (NARES) with centralized hosting of content but decentralized management. The basic idea of open access (OA) policy is to limit the permission barriers for the user and making the content available online without any permission and price barriers. Thus, OA is free, immediate and a permanent online access to research articles for anyone in the world to improve upon the existing research findings. The ICAR adopted the open access policy (http://icar.org.in/en/node/6609) for easy access of information to the community of research, faculty and extension workers. The main points of ICAR, OA policy are:

- Each ICAR institute to setup an Open Access Institutional Repository.
- ICAR shall setup a central harvester to harvest the metadata and full-text of all the records from all the OA repositories of the ICAR institutes for one stop access to all the agricultural knowledge generated in ICAR.
- All the meta-data and other information of the institutional repositories are copyrighted with the ICAR. These are licensed for use, re-use and sharing for academic and research purposes. Commercial and other reuse requires written permission.
- All publications viz., research articles, popular articles, monographs, catalogues, conference proceedings, success stories, case studies, annual reports, newsletters, pamphlets, brochures, bulletins, summary of the completed projects, speeches, and other grey literatures available with the institutes to be placed under Open Access.
- The institutes are free to place their unpublished reports in their open access repository. They are encouraged to share their works in public repositories like YouTube and social networking sites like Facebook ®, Google+, etc. along with appropriate disclaimer.
- The authors of the scholarly articles produced from the research conducted at the ICAR institutes have to deposit immediately the final authors version manuscripts of papers accepted for publication (pre-prints and post-prints) in the institute's Open Access repository.
- Scientists and other research personnel of the ICAR working in all ICAR institutes or elsewhere are encouraged to publish their research work with publishers which allow self- archiving in Open Access Institutional Repositories.
- The authors of the scholarly literature produced from the research funded in whole or part by the ICAR or by other Public Funds at ICAR establishments are required to deposit the final version of the author's peer-reviewed manuscript in the ICAR institute's Open Access Institutional Repository.

- Scientists are advised to mention the ICAR's Open Access policy while signing the copyright agreements with the publishers and the embargo, if any, should not be later than 12 months.
- M.Sc. and Ph.D. thesis/dissertations (full contents) and summary of completed research projects to be deposited in the institutes open access repository after completion of the work. The metadata (e.g., title, abstract, authors, publisher, etc.) be freely accessible from the time of deposition of the content and their free unrestricted use through Open Access can be made after an embargo period not more than 12 months.
- All the journals published by the ICAR have been made Open Access. Journals, conference proceedings and other scholarly literature published with the financial support from ICAR to the professional societies and others, to be made Open.
- The documents having material to be patented or commercialised, or where the promulgations would infringe a legal commitment by the institute and/or the author, may not be included in institute's Open Access repository. However, the ICAR scientists and staff as authors of the commercial books may negotiate with the publishers to share the same via institutional repositories after a suitable embargo period.

OA is a process and expects full compliance over a period of time. Therefore, the OA policy is a first step in the journey towards formal declaration of openness in the system.

## 2. Why Open Access Institutional Repository Needed?

Institutional repository is a "Digital Collection that captures, preserves, archives and provides policy based access to the intellectual output of an institution". It can be perceived as an organization based set of services which the organization offers to the members of its community for the management and dissemination of digital materials created by the institution and its community members. Researchers, faculty and authors in quest for greater impact of their intellectual work share their hard work in the form of research papers, technical bulletins, books etc., with commercial publishers and they don't have to look for the commercial income from it. Their interest is wide dissemination and further follow-up of their research output. On the other hand publishers owing to commercial interest put high subscription cost, thus, restricting the circulation. This creates an impact barrier. On the other hand, researchers, faculty and scientific workers look for easy access to relevant scientific and other literature but do not have easy access to most of the literature for want of monetary cost required to be paid to publishers. This leads to creation of an access barrier. These structural problems with scholarly publishing can be addressed to great extent by creating Open Access Digital Repositories.

For an institutions, an open access digital repository can go long way in raising the profile and prestige of the institution, provide efficient management of institutional information assets, accreditation and performance management.

In long-term such digitally accessible organizational knowledge repository results in cost savings. For the research community it provides an alternative route to free research communication process and helps avoids time lag and unnecessary duplication. However, there are some concerns with respect to quality control - particularly peer review, IPR and copyright issues, etc. In fact, if institutional repository is seen as complementary to the commercial publishing not intended to totally replace it, it can help and advice on IPR issues as outputs are easily available in digitally searchable form. The open access institutional repository can help and advice on formulation of further research strategy for acquisition of relevant knowledge to meet the goals of an organization.

## 3. Importance of Digitization and Preservation

As we are aware, the digital world is a binary world where all the information is represented as bits, 0s and 1s as against conventional analog representation where infinitely variable nature of information is preserved. Contents in every format and medium held by library, manuscripts to maps, moving images to musical recordings can be converted to suitable digital format. In addition to hardware, software needed for conversion and creation of digital content, the required practices for describing the digital content and their retrieval are also developing fast. The advantages of digital content normally outweigh the loss of fidelity in converting the information from continuous to discrete form due to following advantages:

- Wider and easy policy based access,
- Easy presentation and maneuvering of data,
- Compression of large storage space,
- Fast, multi-dimensional and semantic retrieval,
- Reproducibility and repackaging in different forms,
- Simultaneous & endless reusability,
- Digital resources are best for facilitating access to information but questionable when it comes to traditional library role of authentic preservation.
- Digital content is machine-legible only whereas conventional content is eye-legible.
- Digitized information needs computer hardware and software which are often proprietary and become obsolete very fast, requiring conversion to newer formats and technologies.
- Transition from one file format to other may not produce exactly same file, although, there may not be any loss of intellectual content.
- Assuring integrity of digital file and keeping track of versions is another challenge.
- Active human intervention for refreshing and migration of data is required for maintaining it in the fast changing digital technology environment.

- The life expectancy of digital media, the quality of its manufacturing, the number of times the media is accessed over its lifetime, the quality of the device used to write to or read from the media are matter of concern and need careful media handling, storage temperature, humidity and cleanliness of the storage environment.
- DNA-storage: The researchers of the Chinese University of Hong Kong used encoded *E. Coli* plasmid DNA (a molecule of DNA usually present in bacteria that replicate independently of chromosomal DNA) to encrypt the data and store it in the bacteria. Then, by using *a novel information processing system* they were able to reconstruct and recover the data with error checking. Based on the procedures tested, they estimate the ability to store about 900 terabytes (TB) in one gram of bacteria cells. That is the equivalent of 450 hard drives, each with the capacity of 2 terabytes (2000 GB). Another advantage of the system is that the bacteria cells abundantly replicate the data storage units thereby ensuring the integrity and permanence of the data by redundancy. Genetic codes have been preserved using similar strategy by nature. (http://2010.igem.org/Team:Hong_Kong-CUHK/Project)

## 4. Need of Institutional Repository (IR)

Institutional repository is a Digital Collection of information which captures, preserves, archives and provides policy based access to the intellectual output of an institution. It can be perceived as an organization based set of services which the organization offers to the members of its community for the management and dissemination of digital materials created by the institution and its community members. IR helps in increased control by scholars and the academy over the system of scholarly publishing. IR provides scholarly information free of cost or at fair and reasonable price. However, IR should be seen as complementary channel, not intended to replace commercial publishing. The IR increases the visibility and citation impact of an institution's intellectual output and provides unified access to an institution's output. The IR act as digital platform to preserve institution's intellectual assets and help in providing and managing open access to institution's intellectual assets.

Taking the clue from this, strategy of 'replication and evolution' can help manage data preservation, even on magnetic media of today for very long time. Indian National Agricultural Research & Education System (NARES) has a very large collection of conventional knowledge base in agriculture and allied sciences, spread over the country in Institutes and State Agricultural Universities. Digitization of these valuable archives would allow online access to researchers, teachers and students to which they would not otherwise have an easy access. Therefore, Indian Council of Agricultural Research (ICAR) took several digital initiatives to capture and manage knowledge in NARES, one of the largest agricultural research & education system in world. These initiatives came in the form of several subprojects under the World Bank supported National Agricultural Technology Project (NATP) and National Agricultural Innovation Project (NAIP), a major initiatives to reform the way research and development is done in our traditional system. CeRA, E-Granth, Rice Knowledge Management Portal (RKMP), Agroweb, Agripedia, MIS/FMS, ICAR journal portal, supercomputing for Bioinformatics, Computing facilities, etc. are the few impacts making initiatives on digital access has been started by ICAR[NAIP, 2014].

## 4.1. KrishiKosh a Digital Repository of NARES

Indian National Agricultural Research & Education System (NARES) is a huge repository of knowledge and information on crop sciences, horticulture, resource management, animal sciences, agricultural engineering, fisheries, agricultural extension and agricultural education. Digital technologies and online access to information resources have brought increased expectation from library and information services. For researchers, fast access to existing scientific outputs and archived scholarly information on the topic of interest is as crucial as current scientific knowledge. The modes of services that librarians and information professionals provide has thus become very important and have undergone fundamental changes over past few decades. Digital resources, digital services and access technologies continue to create new opportunities, new challenges and new expectations. Union catalogue, digital repository and digital libraries are the new paradigms which have been taken up to facilitate researchers, teachers, students, extension professionals. ICAR also has declared adoption of open access policy for proper utilization of Intuitional knowledge. It has been observed that in the recent years subscription to journals by libraries of ICAR Institutes / State Agricultural Universities (SAUs) has been on the decline mainly because of the increase in the cost of reputed relevant journals and books coupled with reducing fund availability for the purpose. At the same time, the research/educational activities must always keep pace with the international competition for which all important journals and books should be made available to researchers/teachers in the NARES. Maintaining a traditional form of library with hardcopies is becoming labour-intensive and adds to the cost. Each and every library cannot be sustained without adequate funds. NARES must take advantages of sweeping changes taking place globally. Considering these facts the importance of digital repository and digital library under e-Granth becomes more relevant. The institutional repository can hold all the intellectual outputs of the NARES system in the form of digitized institutional publications, technical reports, annual reports, lectures, authors collection in the form of preprints, reprints etc. These contents to which one can easily have open access, essentially captures all the intellectual work being done under NARES. The same intellectual output when gets published in the form of research papers in the commercial journals become inaccessible due to high cost. Thus institutional repository provides alternative source of scientific information to support our quality research and teaching. KrishiKosh is available at http://krishikosh.egranth.ac.in and provides open access to most of its content.

KrishiKosh is a versatile open access digital repository catering to the needs of NARES and has architecture of centralized hosting of content but decentralized management. The KrishiKosh is hosted at the data center of Indian Agricultural Research Institute (IARI), the premier research institute and deemed university under NARES. Each institute or university can manage and administer its own repository which is integral part of KrishiKosh. The KrishiKosh has been designed by using open source software DSpace which has been suitability configured to meet the requirements of NARES. Each institution in NARES has been configured as community in DSpace having its own collections and logo. Each community and collection can be given independent rights to registered users for uploading and managing the contents. Thus, KrishiKosh is a collectively managed, centrally aggregated repository with integrated search facility. The major objectives of KrishiKosh are to create a digital Institutional Repository of important institutional publications including rare books and old journals and make them open access under NARES. The need for improving accessibility coupled with preservation is necessity for implementation of Krishikosh under E-Granth. To create dependable digital storage and an efficient Integrated Content Management System (ICMS), an open source software DSpace has been customized to meet the requirements. It provides following functionalities:

- **Improve Accessibility: -** The ICMS makes the holdings more accessible to scholars, teachers, academics and the general public, both within the premises as well as to those who cannot personally visit the NARES libraries but want to access the contents through the internet, under open access policy.
- **Enhanced Search ability: -** All holdings are grouped communities and collections based on institutions, subjects, themes or other criteria making large amount of information easily available on any subject matter for teaching, research and development. Any researcher looking for content on any subject or themes can have a unified access to content on all media types (manuscripts, photographs, audio-video, etc.) thereby making the searching much easier and faster.
- **Preservation: -** Preservation of all the rare documents in electronic form is an important objective. Also, once the documents are scanned and digitized, preservation of the originals can be ensured for a much longer period as the need to handle the physical documents is eliminated or minimized to a great extent since documents are made available through the ICMS.
- **Content Selection: -** High power committees of subject matter experts have identified the content of intellectual and academic value to be

included in the repository. Other institutions have identified the content in consultation with subject matter experts approved by the Directors/ Vice-Chancellors. The identified content was then harmonized centrally to avoid duplication.

- Various types of archival material at NARES comprises of rare books, old journals, reports, newsletters, annual reports, success stories, special bulletins, convocation addresses, endowment lectures, author's collections, preprints, reprints, patents, manuscripts, periodicals, grey literature, photographs, existing digital content, audio-video recordings.
- It is NARES's intention to make the Metadata for all records (and categories) freely available to all, however the actual records would be accessible based upon its access category.
- All of NARES's holdings are classified under the following three access categories:
- **Public Access** : Any record that can be made available to public at large shall fall under this category
- **Privileged Access:** Records classified under this category shall be accessible to only to those individuals or organizations that have a privileged status with NARES (such as other national / state archives / research and academic institutes / eminent researchers etc.). Others (the world at large) would have to seek prior permission / approval from IARI to access any Record classified as Privileged Access.
- **Prohibited Access:** Records which are accessible ONLY to NARES authorized officials, due to their confidential and sensitive nature as defined by statutory rules and regulation.

Thus, KrishiKosh is a digital repository which captures, preserves, archives and provides policy based access to the intellectual output of Indian NARES. It is a unique repository of knowledge in agriculture and allied sciences, having collection of old and valuable books, institutional publications, technical bulletins, project reports, lectures, preprints, reprints, thesis, records and various documents spread all over the country in different libraries of Research Institutions and State Agricultural Universities (SAUs). The home page of this repository is given below and can be visited through the link (http://krishikosh.egranth.ac.in/). At present KrishiKosh has more than 16 million digitized pages in more than 71,000 digital items (volumes) like old books, old Journals, reports, proceedings, reprint, research highlights, training manuals, historical records. More than 26,000 thesis are submitted at Krishikosh by various SAUs / Institutions and value addition has been done by making theses full Text searchable.

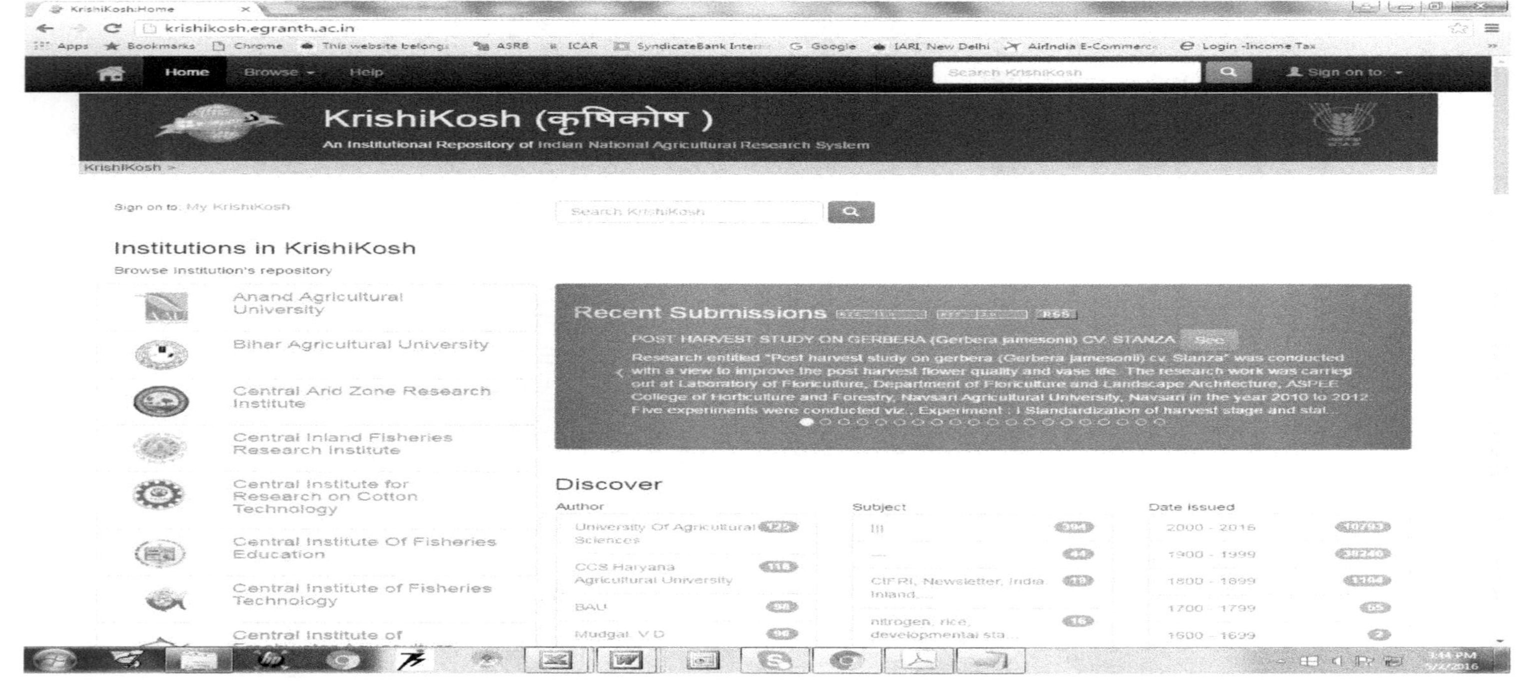
KrishiKosh:Home
krishikosh.egranth.ac.in
Home
Browse
Help
Sign on to:
KrishiKosh (कृषिकोष )
An Institutional Repository of Indian National Agricultural Research System
KrishiKosh >
Sign on to: My KrishiKosh
Institutions in KrishiKosh
Browse institution's repository
Anand Agricultural University
Bihar Agricultural University
Central Arid Zone Research Institute
Central Inland Fisheries Research Institute
Central Institute for Research on Cotton Technology
Central Institute Of Fisheries Education
Central Institute of Fisheries Technology
Central Institute of
Recent Submissions
POST HARVEST STUDY ON GERBERA (Gerbera jamesonii) CV. STANZA
Research entitled "Post harvest study on gerbera (Gerbera jamesonii) cv. Stanza" was conducted with a view to improve the post harvest flower quality and vase life. The research work was carried out at Laboratory of Floriculture, Department of Floriculture and Landscape Architecture, ASPEE College of Horticulture and Forestry, Navsari Agricultural University, Navsari in the year 2010 to 2012. Five experiments were conducted viz., Experiment : I Standardization of harvest stage and stat...
Discover
Author
Subject
Date issued
University Of Agricultural Sciences
CCS Haryana Agricultural University
BAU
Mudgal, V D
CIFRI, Newsletter, India Inland,...
nitrogen, rice, developmental sta...
2000 - 2016
1900 - 1999
1800 - 1899
1700 - 1799
1600 - 1699

Krishikosh platform is an Institutional Repository for collecting, preserving, and disseminating information in digital form for the intellectual output of an institution. In this repository, some important terminology such as Community, Sub Community, Collection, Item and Bitstream needs to be understood. The explanations of these terminologies are given below.

- **Community:** Community is the top level reference term which describes the University/ICAR Institute group. Generally the right to create a Community is with the Administrator of the Krishikosh.
- **Sub Community:** This is second level of hierarchy. It may describe departments/ division under the University/ICAR Institute.
- **Collection:** Collection is a part of Community or Sub-community in which we can add different categories like books, thesis, journals, newsletters etc. Creating collection is necessary to post the document under Krishikosh.
- **Item:** The record/document which is uploaded in collections is termed as item.
- **Bitstream:** It is the file which will be uploaded in the Krishikosh preferably a searchable pdf/a or pdf file.

## 4.2. Integrated Digital Ensemble of Agricultural Libraries (IDEAL)

To strengthen the digital library initiatives, more advanced Library Management Software, compliant to open international standards is necessary for easy data portability and data sharing. Koha is open source software which has been identified for implementation in the NARES libraries with expert support, intensive trainings. In-house capacity building has been part of the strategy. All further advanced library automation system like RFID for automated library services can be built only on robust Library Management System compliant to international standards for data compatibility and portability. Koha has been implemented in the 37 libraries under NARES. Koha is a full featured Integrated Library System (ILS), downloadable free under GNU General Public License, maintained by a dedicated team of software providers and library technology Koha OPAC page at IARI, New Delhi professionals from around the globe. Authorized user may modify the codes to adapt it to local needs and redistribute it. Koha has robust Cataloging, Circulation, Patrons, Search, Serials control, Acquisition, Reports and Administration modules along with utility Tools and OPAC. By adopting it, the customer becomes 'joint stake holder' in the product. Koha is well tried and tested software and has demonstrated both stability and scalability, is being used in hundreds of libraries worldwide. It is an example of Collaboration and Resource Sharing. Software solutions are freely available to all libraries worldwide. Libraries benefits from the contributions of other participating library systems. Being an open source software it has benefit of being free from vendor lock-in whereas, in proprietary software, source code is 'closed' and support and future development of the product solely rely on the success and resources of a the single vendor. If the vendor goes down or does not cooperate, your product support is gone. Open source solutions rely on stable code bases developed and supported by many providers worldwide. Koha is compatible with existing technologies viz. RFID, and being open source developing software, compatibility with any new Library Technology will be available in future also. The IDEAL platform has been developed on Software as Service (SaaS) architecture with independently configured instance for each library having OPAC and staff clients running on centralized servers along with option to run local mirror for individual library. It can be accessed at http://ideal.egranth.ac.in. All the libraries of NARES can join IDEAL platform to get themselves integrated to Virtual Digital Library of NARES and get freedom from maintaining hardware and software locally for day to day functioning of their own library. This enables them to focus more on their core competency of managing their library more efficiently [Jain *et. al.*, 2014; Jain *et. al.*, 2016].

## 4.3. Integrated National Agricultural Resources Information System

INARIS was taken up as a sub-project under National Agricultural Technology Project (NATP). The goal for this project was to design and develop a flexible Central Data Warehouse (CDW) of agricultural resources and databases on different subjects. The target users of information systems and decision support system developed under this project are

i. Research Managers

ii. Research Scientists

iii. General Users.

In this project a state of art Central Data Warehouse (CDW) of agricultural resources of the country has been developed at ICAR-IASRI, New Delhi. This provides systematic and periodic information to research scientists, planners, decision makers and developmental agencies in the form of On-line Analytical Processing (OLAP) decision support system. It has been implemented with active collaboration and support from 13 other ICAR institutions, namely NBSSLUP Nagpur (for soil resources), CRIDA Hyderabad (for agro-meteorology), PDCSR Modipuram (for crops and cropping systems), NBAGR Karnal (for livestock resources), NBFGR Lucknow (for fish resources), NBPGR New Delhi (for plant genetic resources), NCAP New Delhi (for socio-economic resources), CIAE Bhopal (for agricultural implements and machinery), CPCRI Kasargod (for plantation crops), IISR Calicut (for spices crops), ICAR Research Complex for Eastern Region Patna (for water resources), NRC-AF Jhansi (for agro forestry) and IIHR Bangalore (for horticultural crops). In all 59 databases on agricultural technologies generated by council, research projects in operation and related agricultural statistics from published official sources at least from the year 1990 onwards at the district level were integrated into this information system. Subject-wise data marts were created; multi-dimensional data cubes have been developed and published on Internet/ Intranet. The validation checks have been implemented wherever possible. The information of this data warehouse are available to user in the form of decision support system in which the all the flexibility of the presentation of the information, it's on line analysis including graphic is inbuilt in to the system. The system also provides facility of spatial analysis of the data through web using functionalities of Geographic Information System (GIS). Apart from this, subject wise information system has been developed for the general users. The user of this system has the access of subject wise dynamic reports through web. The facilities of data mining and generation of ad-hoc querying were also extended to limited users. Therefore, the dissemination of information from this data warehouse for different categories of users is through web browser with proper authentication of the users. The web site of the project is already launched (www.inaris.gen.in) and the multidimensional cubes, dynamic reports, GIS maps and information systems are already available to the users. This project is viewed to strengthen the information system conceptualized by ICAR. Other agencies, in particular, the planning portfolio, are eagerly waiting for such a decision support system. Based on the interaction among the basic resources like soil, water, climate, animal and vegetation that form the prime components of the production system this data warehouse will help in determining the carrying capacity of the region. The project aims at giving suitable opportunity on multi-disciplinary mode through enhanced linkages among research institutes and other development agencies by providing first hand information on problems and potential in production systems. This data warehouse may be intensively used with an ultimate aim of enhancing better quality of life of the farming community and society at large

## 4.4. Establishment of an Online System for NET/ARS - Prelim Examination

A state-of-the - art infrastructure facility for conducting examination of ARS/ NET of ASRB was created under the subproject with the major objective to develop the capability to change over from on-site to on-line Examination for NET/ARS Prelim. The on-line examination network consisting of one Data Center (DC) at the ASRB, One Disaster Recovery (DR) site near its premises, and 23 Examination Centers (or Nodal Centers) across the country was set up. These Examination Centers were created at 21 ICAR Institutes and two SAUs considering that the management control would be better at these locations being part of the NARES family. These centers were equipped with necessary hardware and software that was developed and customized as per the requirements of the ASRB.

## 4.5. Establishment of Supercomputing Hub for Indian Agriculture

During the last decade, genomics has witnessed an information explosion. Genomic databases contain huge amounts of information that are not amenable to traditional analytical approaches. The analysis of genomic sequences for drawing valid conclusion is highly computer intensive and needs different tools and technique. Apart from this, there is need to design and develop databases and data warehouse of genomic data of local species and commodities to facilitate researchers. Software and web browser based systems need to be developed for visualization, mapping and interpretation of these genomic sequences. Also, there is hardly any consolidated efforts are made for collection, compilation, storage and knowledge mining of indigenous agricultural genomic resources. In order to keep pace with the research and developments in agricultural bioinformatics at global level, country needs expertise and exposure in this area of research. Therefore, there is an urgent need to establish this National Agricultural Bioinformatics Grid

(NABG) which will help in developing databases, data warehouse, software and tools, algorithms, genome browsers and high-end computational facilities through systematic and integrated approach in the field of agricultural bioinformatics. The first supercomputing hub for Indian Agriculture in bioinformatics called ASHOKA (Advanced Super-computing Hub for OMICS Knowledge in Agriculture) was established at the IASRI in New Delhi under the National Agricultural Bio-informatics Grid (NABG) sub-project of NAIP (http://www.nabg.iasri.res.in). The hub consisting of supercomputing systems at the NBAGR, NBPGR, NBFGR, NBAIM and NBAIR constitutes the National Agricultural Bioinformatics Grid in the country.

## 4.6. Development and Maintenance of Rice Knowledge Management Portal (RKMP):

RKMP (http://www.rkmp.co.in) is a technical highway for sharing the knowledge of Rice by using new Information and Mobile Technology. It helps the departments which are working in agricultural activities to reach the farmers through extension advisory services, in the best possible way. This platform are built on Microsoft Web 2.0 technology, it caters to location specific information needs of farmers and research officials through IP based customization on 24X7 bases. RKMP is operating by providing content in local language. RKMP operates in multiple domains e.g. Extension and Farmers that provides production know how, practices, FAQs etc., in local languages and English. In research domain, various services are provided such as AICRIP Intranet, archives of AICRIP data (27000 datasets), communities of practice (CoP), bio-informatics suite, approach papers, India Rice Research Repository (i3R), status papers on rice for different states etc.

The portal works on two e-learning platforms which provide learning opportunity to scientists and extension workers simultaneously. This portal also caters to information needs of exporters and farmers through the trade information system. It also provides indexing of mandi prices of paddy from regulated market yards (from Agmarknet). Policy makers can directly access area, production, productivity trends of last four decades up to district level. In a first kind of an attempt, the users can upload the content as a registered user, irrespective of institutional affiliation. RKMP Nodal officers (AICRIP Scientists from State Agricultural Universities) will validate and approve your content before it is displayed online with due credit to the contributors.

## 4.7. Agroweb-Digital Dissemination System for Indian Agricultural Research (ADDSIAR):

An attempt was made to create a common gateway to ICAR Institutes to act as a one stop window for getting access to all the information about National Agricultural Research and Education System in India. Accordingly, the ADDSIAR was conceived with the broad objective to improve the web presence of ICAR and its Institutes through their websites by making the websites more dynamic and developing a brand image of ICAR. Website Uniformity Guidelines for the ICAR was developed and disseminated which outlined the Standards and Content Management Strategies (CMS) to be employed by all the ICAR Institutes. The ADDSIAR established at the Directorate of Knowledge Management in Agriculture (DKMA) is committed to promote ICT driven technology and information dissemination system for quick, effectual and cost-effective delivery of messages to all the stakeholders in agriculture. Keeping pace with the current knowledge diffusion trends, the Directorate is delivering and showcasing ICAR technologies, policies and other activities through print, electronic and web mode [Tyagi *et. al.*, 2014].

## 4.8. Engaging Farmers; Enriching Knowledge, Agropedia:

Agropedia is a comprehensive and integrated model of digital content organization in the agricultural domain. It aims to bring together a community of practice through an ICT mediated knowledge creating a common platform with an effort to leverage the existing agricultural extension system. There are three groups of agencies/institutions on this project with different roles & responsibilities (http://agropedia.iitk.ac.in/)

*ICT Resource Institutions:*

IIT Kanpur (agropedia platform); IIT Bombay; IIITM Kerala (multi-modal delivery); NAARM

*Agricultural Information and Learning Resources:*

G B Pant University for Agriculture and Technology, Uttarakhand; University of Agricultural Sciences, Dharwad with two KVK's in Maharashtra through the IIT Bombay network.

*ICT4D Interface Partners:*

ICRISAT, with its VASAT project in India and the NAARM will provide the facilitation support for agricultural research scientists and educators and the ICT4D actors. ICRISAT is the consortium leader, which has overall responsibilities for the outputs and deliverables. ICRISAT is uniquely positioned because of its long- standing formal and working relationship with nearly all the partners. Its strength in IT innovations for human development has been described in the IEEE Spectrum (Feb 2004). Its long standing partnership with the FAO (especially in AGROVOC work), and its position as the hosting centre of many CGIAR activities in India, add further strength in implementing this project in a consortium mode. Agropedia has been developed as a common platform for all kinds of information related to Indian agriculture. In one of the first attempts worldwide, the practice of crop knowledge models has been defined and developed to create architecture for accumulating known codified and approved information about crops. This was accomplished with the support of Food and Agriculture Organization (FAO), Rome. Knowledge models (KMs) are the structural representation of knowledge by using symbols to represent pieces of knowledge and relationships between them, which can be used to connect to the knowledge base in agropedia using semantic tools. KMs have been represented using Concept Map (C-Map) tools. KMs have been designed with the intention of using them for indexing and browsing the content that we gather in the repository. A template for objects and relationships within the KMs as well as guidelines to develop KMs were formulated by the NAIP- KM team of IITK with the assistance and support of FAO. Agropedia an agricultural knowledge management portal (http://www.agropedia.iitk.ac.in) was developed as an open platform to facilitate exchange and delivery of information between the agricultural community through a web portal and mobile phone networks.

## 4.9. E-Publishing of Scientific Journals for Indian NARES:

The E-Publishing and Knowledge System in Agricultural Research (EPKSAR) portal developed in the Project has made significant impact on the publishing process and manuscript management of research journals through the implementation of ICT in research journal publishing. Implementation of e-publishing (http://epubs.icar.org.in/ejournal/) has resulted in making the entire publishing process quick, transparent and paperless resulting in the improvement of overall efficiency are being published using the developed ICT enabled platform and are available on-line now. Open Access policy in ICAR for enhanced dissemination and sharing of Indian agricultural research is the outcome of the sub-project.

## 4.10. Consortium for E-Resources in Agriculture (CeRA):

The Consortium for e-Resources in Agriculture, popularly known as CeRA, facilitates online access to about more than 3400 journals in agriculture and allied sciences to all researchers comprising; scientists, teachers, faculty, research fellows and students in the National Agricultural Research System (NARES) through IP authentication. This is the first of its kind for facilitating 24 x 7 on-line accesses of select journals in agricultural and allied sciences to all researchers. At present, there are 147 members (along with regional stations, KVKs and colleges) in CeRA comprising ICAR Institutes, SAUs, NRCs, PDs, etc. in the NARES. About 3,490 journals are now accessible in CeRA [(http://cera.iari.res.in & http://www.jgateplus.com], which is now the most sought after on-line platform by scientists/teachers in the NARES for literature search through IP authentication. The website (http://cera.iari.res.in) has been developed in Joomla platform using PHP, HTML languages in frontend and MySQL database in the backend. Contents in the site comprises general information on CeRA, committees, feedback, available journal lists, information on workshops and important news, important publications, manual, etc. – accessible to general public and information on financial details, agreements with publishers, invoice, unprocessed data, SOEs, etc. – under secured access. In this way, all information of the Consortium is available in one platform. The second website (http://www.jgateplus.com) is the updated version of (http://cera.jccc.in) developed at the time of launch workshop. This site contains metadata of all journals available in CeRA and accessible through IP authentication. Besides, the contact details of each member and the nodal officer of CeRA have to access to generate reports on hits/downloads and DDRS. This is the online platform for access to CeRA journals. Some of the important options facilities available in this platform are: advance search, my favorite journals and live chat with the service provider for online solution to a given problem. The impact of CeRA in research publications is revealed through Web of Science, which indicate qualitative and quantitative increase in the number of published papers during post CeRA (2008-12) than Pre-CeRA (2003-07). CeRA acts like a catalyst to enhance agricultural research, education and extension activities of NARES institutions. This would not have happened but for the constant help and co-operation of all CeRA members all along. The subscription at one place, instead of subscribing individually, provides an efficient way of subscription of research journal under the NARES in terms of time, space and budget [Chandrasekharan *et. al.*, 2012; NAIP (2014)]

## 4.11. Development of E-Courses for degree level programmes in agriculture and its allied areas:

As the traditional methods of educating the new generation of tech-savvy students are found wanting, the need for use of new technologies in agricultural education is gaining momentum. Hence, 425 user-friendly and multimedia-based e-courses for the under-graduate students were developed in seven disciplines viz., agriculture, dairy science, veterinary science and animal husbandry, fisheries science, horticulture, home science, and agricultural engineering comprising 15820 lessons. A dedicated portal on e-courses covering all the seven disciplines was made available at http://ecourses.iasri.res.in, so that the user community could access the desired e-Course contents anytime and anywhere. Off-line DVDs were also distributed to all the SAUs, DUs and other academic institutions in India on demand.

## 4.12. Implementation of Management Information System (MIS) including Financial Management System (FMS) in ICAR

ICAR took an initiative to develop and Enterprise Resources Planning (ERP) solution that will take care of all Institutes and Centers of ICAR as a whole. ICAR-IASRI was identified as a leading center for development and successfully implements MIS (including FMS) System which includes solution for Financial Management, Project Management, Material Management, and Human Resource Management & Payroll at ICAR. Major Benefits of this approach are:

- Centralized data management system across all institutes.
- Finally, creating an IT environment in ICAR across all disciplines.

ERP system was planned under the sub-project entitled Implementation of Management Information System including Financial Management System in ICAR. An ERP system integrates different parts of the business processes and their activities such as planning, purchasing, inventory, sales, project, finance, human resources, etc. Establishment of a Central Data Centre (CDC) of ICAR was further added under the sub-project and the CDC was established at IASRI to address the requirement of MIS-FMS including web hosting and unified messaging solution. The software of ICAR-ERP was developed in the project using Oracle ERP available at http://icarerp.iasri.res.in . The ICAR-ERP solution facilitates efficient and effective planning and management of resources. The system integration processes were carried out in the five major functional areas *viz.* Financial management; Project management; Material management; Human resource and Payroll system. The solution is developed using Java as a driving engine with a backend of Oracle, the system is designed and implemented in centrally and is used by accessing URLs on java enabled Browsers (http://www.iasri.res.in/misfms/ ).

## 4.13. Strengthening Statistical Computing for NARES

Under this sub-project, emphasis was given to strengthen the high end statistical computing environment for the scientists in NARES. Availability of a very healthy statistical computing environment for the scientists in NARES containing a very powerful, all inclusive, a modern, efficient and precise general purpose statistical software package for undertaking a probing, in-depth and accurate analysis of data generated from agricultural research. This is expected to bring a revolution in the analysis of agricultural research data. Exploratory data analysis, which previously was avoided because of non availability of the high end statistical package, would become a common feature of all agricultural research. The power of the package to graphically display dynamic, interactive visual research can enrich the knowledge of agricultural scientists and illuminate concepts which without statistical software package was more difficult to comprehend earlier (http://www.iasri.res.in/sscnars/).

## 5. Discussion

The green revolution in India benefits the livelihood of farmers immensely and enhances agricultural productivity as well. However, there is a demonstrable need for a new revolution may be called digital revolution that may bring the farmers, researchers and policy makers together for smart solutions. In the new digital era with booming mobile, wireless, and Internet technologies, ICT has penetrated even in poor smallholder farms and in their daily activities. The ability of ICTs may become a powerful tool for farmers to access and organize the available knowledge through the digital initiatives taken by ICAR. It facilitates the implementation of technologies –both new and traditional– and transforming patterns of learning and interactive strategies among researchers for real time solutions.

The digital initiatives especially that of E-Granth and CeRA have consistently enhance the quality and quantity of research output in terms of research papers, methodologies and patents. This is because of the fact that all publishers/journals in agricultural sciences are available on CeRA platform and all institutional repository especially thesis are available on E-Granth portal. Thus, these initiatives play a key role in the research and developmental activities in NARES. ICT application based sub-projects have resulted in better and economic access to quality publications for researchers and students. This has greatly impacted the overall quality of the research publication from the NARES. The National Agricultural Bioinformatics Grid (NABG) provides the platform for research and development in agricultural bioinformatics for inter-disciplinary research in cross-species genomics along with the capacity building. It is expected that, in due course of time information and knowledge generated through research on bioinformatics from the genomic knowledge base will start flowing downward to researchers to users and experimentations in different sectors of agriculture can be able to evolve internationally superior competitive varieties/breeds and commodities in agriculture. It is also estimated that the total amount of information doubles every four to five years. ICTs are crucial in coping with the explosion of knowledge in agricultural sciences such as genomics huge data were generated. Supercomputing facility available with advanced statistical tools were utilized to convert these knowledge's into an information.

RKMP is strengthening the research, extension, farmers, private sub-systems, partnerships and networks for the better flow of rice knowledge. It is also providing vital scientific information and contributing to overall rice development in the country. The developed portal helped to strengthen communication infrastructure among the stakeholders, improve tools for collecting data and information, nurture scientific communities in the field of rice, provide platform for collaborative action and information sharing, initiate steps for integrating information systems and improve the knowledge sharing culture throughout various key players and stakeholders in the rice sector. The E-Publishing portal has made significant impact on the publishing process and manuscript management of research journals through the implementation of ICT in research journal publishing and Implementation of e-publishing. ERP system was also developed for Management Information System including Financial Management System in ICAR. An ERP system integrates five major functional areas *viz.* Financial management; Project management; Material management; Human resource and Payroll system e-governance system. The ERP system assists the implementation of policy frameworks and to monitor progress for any organization. For strengthen the high end statistical computing environment for the researcher, faculty and students in NARES, statistical computing environment were provided for efficient and precise general purpose statistical systems for undertaking a probing, in-depth and accurate analysis of data generated from agricultural research. This is expected to bring a revolution in the analysis of agricultural research data for knowledge discovery.

The digital portal provides actionable information to community and policy maker on disaster prevention in real time along with advice on risk mitigation technique. This portal made significant impact in the field of risk management. Through regular and systematic surveillance, disastrous situations can be avoided by detecting the events well in advance. Using these technologies advice on the risk mitigation technique can be generated or communicated to the policy makers (RKMP portal can give information in advance on pests disaster in the rice crop. Internet technologies capture pest information from fields and produce – instant and customized pest reports to the plant protection experts to advise the state agriculture agencies who further advise concerned farmers and the same information is available for agricultural policy planners.)

## 6. Conclusion

ICAR initiated a digital initiative to capture and manage Knowledge in NARES. Under these initiatives, platforms were developed for improving the access to appropriate knowledge which leads to smart solutions. These solution bridges the gap between agricultural researchers, extension agents and farmers hereby enhancing the agricultural output research in terms of quality and quantity.

## References

Chandrasekharan, H., Patle, Sarita, Pandey, P S, Mishra, A K, Jain, A K, Goyal, Shikha, Pandey, Amit, Khemchandani, Usha and Kasrija, Rajkumari (2012). CeRA - the e- journal Consortium for National Agricultural Research System; *Current Science*, 102 (06): 847–851

Jain, A. K., Chandrasekharan, H. and Kumar Rajesh (2014). Final Report of NAIP sub-project 'Digital Library and Information Management under NARES (e-GRANTH)'. Indian Agricultural Research Institute, New Delhi, India

Jain, A. K., Kumar, Amrender, Batra, Kamal, Kapur, Sanjiv (2016). Reference Manual on Krishikosh - A Repository for NARES. ICAR– Indian Agriculture Research Institute, New Delhi, pp50

NAIP (National Agricultural Innovation Project) (2014). An Initiative Towards Innovative Agriculture. Final Report, National Agricultural Innovation Project, ICAR, New Delhi

Tyagi R.K., Agrawal A., Varshney, H., T. Ravisankar, K.V. Rao, H. Chandrasekharan, M.K. Chandra Prakash, G.R.K. Murthy, Jain, D.K., Singh, N., Dhandapani, A. and Agrawal, R.C. (2014) Final Report of NAIP sub-project 'AGROWEB-Digital Dissemination System for Indian Agricultural Research (ADDSIAR)'. National Bureau of Plant Genetic Resources, New Delhi, India

# 6

# Social Media: New Vista in Agricultural Education

**Dr D Thammi Raju**

Principal Scientist
Education Systems Management Division
ICAR- National Academy of Agricultural Research Management
Rajendranagar: Hyderabad

## 1. Introduction

Transformation sweeping teaching – learning process in terms of methods, tools and approaches. Technology Enhanced Learning has influenced the educational institutions and gradually traditional teaching is undergoing metamorphosis. Libraries have been replaced by Google, everything is in a reach of the tip of the finger. Social media reduced the distances between students and teachers; became a part of daily life. A change in the teaching style and style of knowledge dissemination is very much evident. Universities today seldom use and appreciate what students use / interact with social media when it comes to learning (formally / informally). Classroom based lectures are still seen as a source of information / knowledge / feedback, but technology has made rapid changes. Sometimes the restriction on the use of mobile phones, deter them from blending with the classroom environment and gives them a feeling of a parallel world which is very different from the world outside University where they are always engrossed in technology and social media (Nair and Singh, 2013). At the same time skepticism and criticism is prevailing over the use of social media in curriculum; leading to distraction for students hampering their cognitive skills. However, the reluctance surrounding the use of social media was partly attributed to the teachers' belief, time required for preparation of course material, technological updating by oneself etc. Proper balance between distraction and opportunities of use of social media in higher education is desired. Social media has the potential to enhance learning ability and provide platform for interaction among peers, instructors and people around. These are web based tools of electronic communication that allows users to interact, create, share, retrieve and exchange information and ideas in any form (text, pictures, videos etc.) that can be discussed upon, archived and used by anyone in virtual communities and networks.

## 2. Social Constructivism

Blending of social media with technologies and innovative pedagogy strongly supported social constructivism, a concept which portrays that knowledge is something that is constructed with in the social context. People within a community help each other out, lend support, interact with one another, serve as shadow guides, and build on each others' progress. One of the major tenets is that this construction is always collaborative. This is to say, learning is not simply the memorization of information, but it is a situation where people construct their own meaning. Social constructivism is much more relevant in Agricultural Education. The social constructivism, backbone of social media, improves the reliance of formal communication channels. In student-centered learning in the context of agricultural education, learning works best when it takes place within a social context since it is an active and constructive process. Currently paradigm shift in the teaching learning process noticed - memorization of information to learning how to access information. Agricultural Education, per se, is contextual in nature – wide diversity in agro climatic conditions, agro biodiversity, culture, socio economic conditions, land holding, use pattern, rural- urban divide etc. Innovative technologies have the potential to promote active engagement, encourage students to work in groups, provide opportunities for feedback from a wide audience, and connect students to others who are knowledgeable in a host of areas, thus social media plays an important role in Agricultural Education. Innovative technologies transform large class lecture teaching to small group interactions such as flipped classrooms. Traditional education is often teacher focused & lecture based and the student work on mostly individual production thus the ability to "harness the potential of collective intelligence" is lost. All of this suggests that our implementation of social media and innovative technologies into our pedagogy correlates well with social constructivist thinking about learning in general. The traditional teaching demands paying attention to the lecture which takes back to a teacher-dominated environment that does not allow for greater learner autonomy - potential learning is sacrificed for greater control. (Orlando, 2011).

Löfström and Nevgi (2006) provided a list of how the use of innovative technologies applies to constructivist principles:

- Learners construct knowledge as a collaborative activity.
- Learners benefit from the cognitive process of working towards a goal.

- Learners use previous knowledge to build on new knowledge.
- Thinking, emotions and action lead to empowerment, commitment, and responsibility.
- Learners actively and purposely set cognitive objectives.
- Learners collaborate by sharing knowledge with other members of a community, engaging in dialogue and receiving feedback.
- Learners reflect on the process and understand the implications.
- Learners connect learning to the context of the real world and transfer knowledge to new applications.

## 3. Social Media Characteristics

Social Media provides an inexpensive, easy to use, interactive, dynamic and collaborative regulated almost ubiquitous and democratic plat form for webdenizens to voiceopinions, express beliefs, share thoughts, and participate in discussion.

## 4. Characteristics of Social media in learning

- **Accessibility:** Publicly available for almost for free or no cost. Teacher develop and share teaching materials to facilitate collaborative learning.
- **Permanence:** Social media sites are more adaptive and evolving platforms for curriculum development. Teachers and learners can edit their blogs, profile etc. anytime.
- **Reach:** Social media sites are hosted on the Internet that reaches the global audience.
- **Recency:** Time lag between communication is almost zero. It may instantaneous and enables collective learning and collaboration among learners.
- **Usability:** Social media doesn't require any special skills and hence both instructors and learners can use for curriculum development and delivery. **(Aggarwal, 2011)**
- **Connectivity:** Social media is tool for connecting students to share their daily experiences, tweeting, sharing moments, experiences. These daily experiences through reading, writing, watching, could enhance student competencies, that forms the crux of students learning process (Nair and Singh 2013).
- **Active online engagement:** Social media enables users / students actively engage with the shared message, by commenting or retweeting, commenting on pictures, videos, liking etc. These acts invoke active learning among students.
- **Knowledge gaining process:** Social media enables users to stream line their information by participating in virtual communities / groups throughout the day with faster speed, unlike world wide web which is loaded full of generic information on different subject matters with little direction.

- **Digital Literacy:** Social media developed digital competencies to a greater extent, a resultant of which, students are very comfortable in using complex interface enabling themto explore, create and test out their own ideas.
- **Developing new skills:** Social media stimulates group dynamics. For eg game simulations enables students to work as a group and coordinate as a group.
- **Many to One relationship:** Human computer interaction establishes many to one relationships. Students / Instructorsdevelops the ability to interact with many using many platforms/ software in teaching and learning. Teachers also can make learning materials and learning process much more interactive and engagement with the help of social media.
- **Personalised Learning:** Cognitive needs and learning styles do vary among students. Social media enables personalized learning instead of 'one size fits all' followed traditionally. Self-directed and self- paced learning takes place.

## 5. Social Media Technologies

Classification of different types of social media platforms are (Kalpan and Haenlein, 2010):

i. **Social networking sites –** Orkut, Facebook, Friendstar, MySpace, Google +
ii. **Blogs and micro blogs :** Blogger, Wordpress, Twitter, Instagram
iii. **Collaborative projects:** Wikis
iv. **Virtual Social world:** Second life
v. **Social gaming:** World of Warcraft, Farmbook
vi. **Content communities:** Video (Youtube, Vimeo, Vine), Photo ( Instagram, Flickr, Tumbler), Audio ( Soundcloud, Podcasts)

Apart from the above, the following were also categorized under social media( Suchiradipta and Saravanan, 2016)

- **Forums, discussion boards and groups:** Google hangout, Blackbaord, Discussion groups ( Dgroups)
- **Socially integrated messaging platforms:**WhatsApp, Facebook messenger, Snapchat
- **Professional networking:** ReseachGate, Academca.edu, LinkedIn
- **Social News :** Reddit, Propeller, Digg
- **Social bookmarking:** Delicious, Blinklist

## 6. Brief description of some of the commonly usedsocial media tools

**Facebook** is a social utility that connects people with friends and others who work, study and live around them. People use Facebook to keep up with friends, upload an unlimited number of photos, share links and videos, and learn more about the people they meet. Facebook was originally launched in 2004 for Harvard students, and then was expanded to other Ivy League universities. Facebook is made up of six primary components: personal profiles, status updates, networks (geographic regions, schools, companies), groups, applications and fan pages (Rachal)

**MySpace** is a social networking website offering an interactive, user-submitted network of friends, personal profiles, blogs, groups, photos, music, and videos. MySpace allows users to fully customize their profiles by complete changing the appearance, background and format of their page

**Google + (Google Plus)** is an internet -based social network that is owned and operated by Google Inc., where like-minded individuals communicate, collaborate and create as a community. It acts as a "social layer" for Google, the world's largest search engine company, and is integrated into its every product and service, such as Search, Android, YouTube, Chrome, Local and Drive.Google+ has the characteristic features - public user profile, circles, stream, identity service, hangouts and hangouts on-air.

**Blogs or Web logs** are a collection of articles written by people arranged in reverse chronological order. It is a discussion or informational site published on the World Wide These articles are known as blog spots. Collection of all blogs is referred as Blogosphere. Blogs allow people to share their views, express their opinions, interact and discuss with each other through linking to other blogs or posting comments. A blog when maintained by an individual is known as individual blog or when managed by a group of people is known as community blog. Authors of blogs are known as bloggers.

Blogging can be seen as a form of social networking service and bloggers do not only produce content to post on their blogs, but also build social relations with their readers and other bloggers. A typical blog combines text, images, and links to other blogs, web pages, and other media related to its topic. The ability of readers to leave comments in an interactive format is an important contribution to the popularity of many blogs. In education, blogs can be used as instructional resources. These blogs are referred to as edublogs.

**Micro blogging** is a broadcast medium that exists in the form of blogging. A microblog differs from a traditional blog in that its content is typically smaller in both actual and aggregated file size. Microblogs "allow users to exchange small elements of content such as short sentences, individual images, or video links", which may be the major reason for their popularity. These small messages are sometimes called *microposts*.

These sites are similar to blogs except the fact that the articles can only be certain length. In case of Twitter, the articles can be 140 characters in length. These articles also called as messages (or tweets in case of Twitter). These sites are typically used to share what one is doing. Besides posting messages, people can also create friendship networks. They can also become followers of other users.

**Wiki** is a piece of server software that allows users to freely create and edit Web page content using any Web browser. Wiki supports hyperlinks and has a simple text syntax for creating new pages and crosslinks between internal pages on the fly. Like many simple concepts, "open editing" has some profound and subtle effects on Wiki usage. Allowing everyday users to create and edit any page in a Web site is exciting in that it encourages democratic use of the Web and promotes content composition by nontechnical users ( http://wiki.org/wiki.cgi?WhatIsWiki).

These are publicly editable encyclopedias. Anyone can contribute articles to wikis or edit existing ones. However, most of the wikis are moderated to protect content from vandalism. These provide a great technology for content management, where people with a basic knowledge of formatting,contribute and produce rich source of information. Popular wikis are Wikipedia. These wikis are a great example of collective in intelligence (Aggarwal, 2011). Franklin and Thankachan (2013) established that the development of the wiki as a support for problem based online course supported more discussion than face to face courses.

**Social book marking** is a centralized online service which enables users to add, annotate, edit, and share bookmarks of web documents. Many online bookmark management services have launched since 1996; Delicious, founded in 2003, popularized the terms "social bookmarking" and "tagging". Tagging is a significant feature of social bookmarking systems, enabling users to organize their bookmarks in flexible ways and develop shared vocabularies known as folksonomies.

These are also known as collaborative tagging, allowing to tag their favourite webpages or websites and share the results with other users. This generates a good amount of metdata for the webpages. People can search through this metadata to find relevant or most favorite webpages / websites. People can also see most popular tags or the most freshly used tags and freshly favoured website / webpages. Social bookmarking is a great example of collective intelligence.

**Social Friendship Networks** allowpeople to stay in touch with their friends and also create new friends. Individuals create their profile on these sites based on their interests and, locations, education, work and so on.

**Social News** sitesallow people to share news with others, permit others to vote on these stories. News items that are voted the most emerge as the most popular news stories. People can obtain most popular stories, fastest upcoming stories for different time period and share their thoughts by providing comments.

**YouTube** is the online video which allows users to easily upload, view, rate, share, and comment on videos across the Internet through Web sites, mobile devices, blogs, and e-mail. It has become a platform for easy distribution of videos to a much wider audience. It makes use of WebM, H.264/MPEG-4 AVC, and AdobeFlash Video technology to display a wide variety of user-generated and corporate mediavideo. Available content includes video clips, TV clips, music videos, movie trailers, and other content such as video blogging, short original videos, and educational videos. Unregistered users can watch videos, and registered users are permitted to upload an unlimited number of videos and add comments to videos.

**Podcasts** are brief audio or video messages created by an individual or group and readily available on the Internet. Audio podcasts are available at a variety of Web sites, including iTunes. Some podcasts feature enhanced video, meaning the video includes voice recording, music, pictures, and/or animation. Podcasts are useful for demonstrating how to perform a task or sharing essential information. Podcasts can easily be uploaded online for sharing with the global community using video-sharing Web sites. Because users are usually well versed at locating, downloading, and playing videos that are available online, little or no instruction is generally needed. (Xie & Gu, 2007).

**Flickr** is an online photo site where users upload photos that can be organized in sets and collections. Public photos may be viewed and commented on by others. Universities have found Flickr to be a great tool to easily share photos with students, alumni, faculty and staff. The automation of uploading the photos, adding captions and tags, and turning them into organized collections with slideshows without any manual Web coding, is a great timesaver for times trapped Web professionals.

**WhatsApp** is a cross-platform instant messaging application that allows iPhone, BlackBerry, Android, Windows Phone and Nokia smartphone users to exchange text, image, video and audio messages for free. WhatsApp is especially popular with end users who do not have unlimited text messaging. In addition to basic messaging, WhatsApp provides group chat and location sharing options.

**Linkedin** is the world's largest professional network with more than 433 million members in 200 countries and territories around the globe. The Mission is to connect the world's professionals to make them more productive and successful. It helps to get access to people, jobs, news, updates, and insights. Some of the applications are - 'Job Search', 'Lookup', 'Lynda.com', 'SlideShare', 'Groups', 'Pulse', 'Recruiter', 'Sales Navigator', 'Elevate'.

**Twitter** is a mix between instant messaging and blogging that allows users to send short (140-character) updates. Users can also follow the updates of friends they "follow," send them direct messages, reply publicly to friends, or just post questions or comments as their current status.Twitter can be used for creating awareness and branding, promoting content, fast feedback, finding new audiences, and marketing – all areas marketers in higher education should have great interest in. In the education, Warren and Wakfield (2013), suggested to allow use of twitter by students to connect as a social space for added value.

## 7. Potential Areas of use of Social Media in Agricultural Research and Education System

The potential areas of social media application in teaching, research and extension sub systems of National Agricultural Research Systems vary. They are

### 7.1. Eeducation

- Academic Communication
- Sharing of educational resources - lecture notes, presentations etc.
- Sharing of photos
- Sharing of videos links
- Posting of comments

- Networking
- Evaluations of assignments

### 7.2. Research

- Sharing of research communications
- Networking and collaboration
- Development of interface with extension and farmers

### 7.3. Extension

- Dissemination of technologies / products
- Dissemination of Information
- Announcements
- Extension bulletins
- Sharing of photos, videos
- Answering queries
- Timely advices -climate, market etc.
- Networking of farmers

## 8. Framework & Guidelines for Use of Social Media for Government Organizations

The Department of Electronics and Information Technology, Ministry of Communications & Information Technology, Government of India brought out frame work and guidelines for use of social media by Government organizations. Following are the salient features.

Information Communication Technologies (ICTs) including internet and mobile based communications are increasingly becoming pervasive and integral to day-to-day functioning of our lives- whether personal or official. ICTs offer an unprecedented opportunity of connecting to each and every individual and design the communication structure accordingly to each person. While at a personal level, the uptake and usage of such media is gaining rapid popularity, use and utility of such media for official purpose remain ambiguous. In order to encourage and enable government agencies to make use of this dynamic medium of interaction, a Framework and Guidelines for use of Social Media by government agencies in India has been formulated. These guidelines will enable the various agencies to create and implement their own strategy for the use of social media. Briefly, the elements of the framework and associated guidelines are as follows:

The framework comprises of the following 6 elements:

- **Objective:** It is not just to disseminate information but also to public engagement for a meaningful public participation of public policy. Government organisations are exploring the use of social media for public engagements for disseminating information, policy making, recruitment, generating awareness, education etc. about public services.

- **Platform:** While social networks currently seem to be the face of social media, they are not the only platform. Some of the other forms of social media include, Social bookmarking site – stumble upon; transaction based platforms – Amazon & eBay; self-publishing media – You Tube, Picasa; Business management etc. Since the choices are many, it is essential to identify one or two key platforms from which the department may begin interaction. Based on objective and response, the basket of platforms may be enhanced.Government departments and agencies can engage social media in any of the following manner- making use of any of the existing or creating their own communication platforms, owned or externally leveraged by following some of the guidelines – duration, type of consultation, scope, existing laws etc.
- **Governance**: Since the official pages of departments must reflect the official position, some measure of control must be included in the flexible design of communication. Just as rules and regulations exist for interaction with traditional media, similar rules must be created for engaging with social media. Some of the key aspects of such a governance structure include - Account Creation, Response and Responsiveness, Resource Governance, Content Governance, Legal Provisions, Data & Information Security Governance etc.
- **Communication Strategy:** Social media can only be used by the Government to communicate existing Government information and propagate official policy to the public. Specific, tailored and verified content to be published and it is one of the components of the overall citizen engagement strategy.
- **Pilot**: Since social media are relatively new forms of communication, it is always better to test efficiency and efficacy of such an initiative with a pilot project. Some of principles of creating such a pilot are – Focused objective setting, begin small, multiplicity of access, content management, community creation
- **Institutionalization:** Properly integrated with the existing administrative and communication structure. An indicative list includes: rules may be established that all policy announcements will be undertaken simultaneously on traditional as well as social media; all important occasions as far as possible may be broadcasted using social media; all documents seeking public opinion must be posted on social media sites; all updates from the website would automatically be updated on social media sites and; all traditional communications will publicize the social media presence.

## 9. Core Values for using Social Media

Using of social media for official purposes, the following points to be kept in mind to smoothen interaction

- **Identity**: Always identify clearly who you are, what is your role in the department and publish in the first person. Disclaimer may be used when appropriate
- **Authority**: Do not comment and respond unless authorized to do so especially in the matters that are sub-judice, draft legislations or relating to other individuals.
- **Relevance**: Comment on issues relevant to your area and make relevant and pertinent comments. This will make conversation productive and help take it to its logical conclusion.
- **Professionalism**: Be Polite, Be Discrete and Be Respectful to all and do not make personal comments for or against any individuals or agencies. Also, professional discussions should not be politicized
- **Openness**: Be open to comments – whether positive or negative. It is NOT necessary to respond to each and every comment
- **Compliance**: Be compliant to relevant rules and regulations. Do not infringe upon IPR, copyright of others
- **Privacy**: Do not reveal personal information about other individuals as well as do not publish your own private and personal details unless you wish for them to be made public to be used by others

## 10. Conclusions

Social media literacy for students and instructors are essential. Synchronous and asynchronous learning is the order of the day. Acquisition of knowledge on different subjects or contemporary issues; not often dealt in university curriculum, has become essential for students and instructors to face the challenges. We are no longer just consumers of media, but content creators and distributors, as well as editors, opinion makers, and journalists. Course design is a continuous process in which modules or components are revised or redesigned continuously to provide effective learning outcomes. Student participation, engagement and understanding are three pillars for such a continuous revision or redesign. These students are no longer passive learners simply reading, and repeating; instead they interact actively with faculty and respond to the learning content as participants in the pedagogical process.

Due to advancement of instructional methods and technologies, learning takes place not only through face to face but also increasingly through technology mediated learning. A higher education instructor need to develop a learning landscape for his students harness many discoveries such as social media in their courses as there is powerful linkage among speech, social experience, scaffolding, learning teaching. The higher education instructor should also examine how the incorporation of blogging and social media technologies can create a rapid, responsive, peer and scaffolder environment for both learning and teaching by employing specific types of communication actions. By utilizing social media channels for engaging students and encouraging particular students in particular communications, the teacher could amplify regular class room discussion, connectedness and interactivity leading to quality agricultural education.

## References

Agarwal N (2011) Collective learning: An integrated use of social media in learning environment. *In* Social Media tools and platforms in learning environments, Edited by White B, King I, Tsang P, Spinger- Verla Berlin Heidelberg. Pp 37–52

Franklin T and Thankachan (2013) Developing a Wiki for problem based online instruction and web 2.0 exploration. *In* Using Social Media effectively in the classroom- Blogs, Wikis, Twitter, and More; Edited by Kay Kyeong- Ju So, (2013) Routledge, New York pp 98-114. https://en.wikipedia.org/wiki/Blog

Joanne Kinsey (2010) Five Social Media Tools for the Extension Toolbox, Journal of Extension, Volume 48 Number 5, Article Number 5TOT7

Löfström, E., & Nevgi, A. (2006). From strategic planning to meaningful learning: Diverse perspectives on the development of web-based teaching and learning in higher education. *British Journal of Educational Technology, 38,* 312–324.

Nair, Uday and Singh P (2013) Food for thought: Can Social Media be a potential 'Learning Tool' for Universities?. Education Quest 4(2), August 2013, pp 111-115.

Orlando R. Kelm (2011) SOCIAL MEDIA: IT'S WHAT STUDENTS DO, *University of Texas at AustinBusiness Communication Quarterly*, Volume 74, Number 4, pp 505-520. DOI: 10.1177/1080569911423960

Rachel Reuben ,The Use of Social Media in Higher Education for Marketing and Communications: A Guide for Professionals in Higher Education.http://www.fullerton.edu/technologyservices/_resources/pdfs/social-media-in-higher-education.pdf

Suchiradipta B and Saravanan R (2016) Social Media: Shaping the future of agricultural extension and advisory services, GFRAS interest group on ICT4RAS discussion paper, GFRAS: Lindau, Switzerland

Warren SJ and Wakfield J S (2013) Learning and Teaching as Communicative actions: Social Media as Educational tool. *In* Using Social Media effectively in the classroom- Blogs, Wikis, Twitter, and More; Edited by Kay Kyeong- Ju So, (2013) Routledge, New York pp 98–114

Wikipedia (2016) https://en.wikipedia.org/wiki/Main_Page

Xie, K., &Gu, M. (2007). *Advancing Cooperative Extension with podcast technology. Journal of Extension,* [Online], 45(5) Article 5TOT2. Available at: http://www.joe.org/joe/2007october/tt2.php

# 7

# A.K. Jain Alias Agricultural Knowledge: Journals–Agricat–Institutional Repository–Networking

**Dr.G.Rathinasabapathy, Ph.D.,**

University Librarian

Tamil Nadu Veterinary and Animal Sciences University

Chennai – 600 051, Tamil Nadu

*e-mail: librarian@tanuvas.org.in*

## 1. Agricultural Knowledge

Agricultural Knowledge needs to be transferred from the global knowledge base and from local research to farmers, enabling them to clarify their own goals and possibilities, educating them on how to make better decisions, and stimulating desirable agricultural development. Effective and efficient dissemination of agricultural knowledge to the stakeholders using the modern science and technology tools will ensure better agricultural growth so that we can reduce hunger and poverty, improve rural livelihoods, and facilitate equitable, environmentally, socially and economically sustainable development.

Agricultural Knowledge resides in many places in many formats *viz.*, journals, books, theses, dissertations, reports, conference proceedings, databases, datasets, etc.

Here, we should not forget that agricultural knowledge is also residing in individuals like scientists and farmers. However, the scientific community prefer documenting knowledge through journals due to various reasons. So, the stakeholders of agriculture should use journals to get latest knowledge. Considering the significance of journal literature, efforts were taken by the e-Granth team led by Dr.A.K.Jain to identify and digitize rare and useful journals kept in various agricultural libraries across the country. In fact, this effort is one of the major digitization initiatives taken up in India. Four regional digitization units were established to cover the entire NARES libraries.

## 2. Journals

Journals play a vital role in scientific communication and without quality journals, quality education and research are not possible. Though the Consortium for e-Resources in Agriculture (CeRA) provides access to latest journals with limited access to back files, access to old back issues of journals was a major bottleneck to the research scholars and scientists of NARES. Since the back files of journals were available in many SAU and ICAR Institute libraries, it was very difficult for the research scholars and scientists to access them. Since this issue was fully aware of by the e-Granth project team, it was proposed to digitize the old back files of journals available at various Agricultural libraries.

Dr.A.K.Jain has opened doors not only for the project partners but also for any agricultural libraries in the country. He permitted the libraries which are out of the umbrella of e-Granth project also to submit journals and old documents for digitization under this project. His magnanimity paved the way for mass digitization of very old and valuable journal back files and gray literature kept in various libraries. Today, millions of pages of digitized journal articles **(17,402 journal volumes)** are available online through **Krishikosh** and the details are furnished in Table-1.

**Table-1: Details of Journals Digitized by Top 10 Partner Libraries**

| Sl. No. | University | Journals |
|---|---|---|
| 1. | UAS, Bengaluru | 6011 |
| 2. | IARI | 5431 |
| 3. | IVRI, Izatnagar | 2318 |
| 4. | IVRI, Mukteswar | 1015 |
| 5. | PJTSAU | 999 |
| 6. | TNAU | 364 |
| 7. | ICAR | 284 |
| 8. | CMFRI | 242 |
| 9. | TANUVAS | 154 |
| 10. | CCSHAU | 125 |
| | Total | 16943 |

## 3. AGRICAT

Though there were many big libraries in the NARES system, they were functioning without any cooperation and there was not even any linkage between these resourceful libraries. Realising the importance of a national level union catalogue so as to ensure single point access to the invaluable information resources built in these special libraries of State Agricultural Universities and ICAR Institutes, the e-Granth project initiated the 'Agricat" in which Dr.A.K.Jain played a vital role.

Agricat is a Union Catalogue of the holdings of 12 major libraries viz., Indian Agricultural Research Institute (the National Agricultural Library), Indian Veterinary Research Institute (National Veterinary Library), UAS-Bangalore, GB Pant University of Agriculture & Technology, CCS Haryana Agriculture University, ANG Ranga Agricultural University, National Dairy Research Institute (National Dairy Science Library), Karnal; CSK Himachal Pradesh Krishi Vishvavidyalaya, Palampur; Mahatma Phule Krishi Vivishvavidyalaya, Central Institute of Fisheries Education, Mumbai (National Fishery Science Library); Tamil Nadu Veterinary and Animal Sciences University, Chennai; and ICAR Headquarters-DKMA, New Delhi.

The Agricat enables the stakeholders of agriculture viz., students, research scholars, faculty, scientists, policy makers and agriculturists to have single point access through the Agricat website to the invaluable resources available in the above libraries. Though there are many regulatory bodies like ICAR are functioning in India such as Medical Council of India (MCI), All India Council of Technical Education (AICTE), etc., there is no such national level online union catalogue available to other disciples. The completion of Agricat project was a herculean task as it involved a dozen libraries and libraries spread across the country but Dr.A.K.Jain has very successfully completed the Agricat in a record time which vouches his leadership quality.

## 4. Institutional Repository

Institutional Repository is an online archive of an institution's scholarly papers, deposited by their authors. Institutional Repositories may include a variety of research output of an organization, such as datasets, administrative documents, course notes, learning objects and conference proceedings. An institutional repository is a means to ensure that the published work of scholars is available to the academic community even after increases in subscription fees or budget cuts within libraries. Institutional Repositories provide scholars with a common platform so that everyone in the institution can contribute scholarly material to promote cross-campus interdisciplinary research. An institutional repository is an online archive for collecting, preserving, and disseminating digital copies of the intellectual output of an institution, particularly a research institution.

### 4.1. Krishikosh

The need for an Institutional Repository for all major institutes and agricultural universities in India was felt by the Indian Council of Agricultural Research (ICAR), an autonomous organisation under the Department of Agricultural Research and Education (DARE), Ministry of Agriculture and Farmers Welfare, Government of India, New Delhi.

Therefore, the ICAR has taken the initiative "Strengthening of Digital Library and Information Management under NARES (E-Granth)" a sub-project under National Agricultural Innovation Project (NAIP) to provide digital access to different library resources for National Agriculture Education and Research System in India which comprises of Agriculture Research Institutions and State Agricultural Universities functioning in India. In this initiative, a digital repository "Krishikosh" (http://krishikosh.egranth.ac.in) has been implemented for the entire NARES. Considering the importance of an Institutional Repository for the entire NARES, the e-Granth team led by Dr.A.K.Jain initiated a common Institutional Repository for the NARES and it is named as "KrishiKosh". The screen shot of Krishikosh web site is depicted in Figure-1.

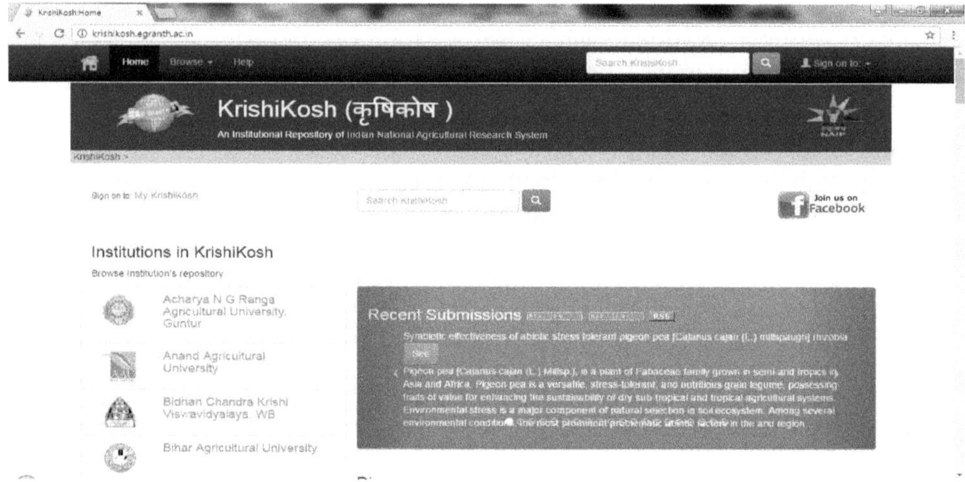


**Figure-1: Screen shost of Krishikosh**

URL: http://krishikosh.egranth.ac.in

In the initial phase, mass digitization was carried out by the IARI with four national level digitization centres and all those interested State Agricultural Universities (SAUs) and ICAR Institutes were given opportunity to get their print documents digitized under this project.

Under the ICAR's Open Access policy, Krishikosh provides ready software platform to implement all aspects of the open access policy, similar to 'Cloud Service' for individual institution's self-managed repository with central integration. IARI, New Delhi is hosting the 'Krishikosh' Repository at AKMU Data Centre of IARI. For easy access to this repository, new initiatives have been taken up for efficient development and dissemination of Krishikosh to the end users of NARES. These initiatives includes the efforts for addition of new library, development of tutorial video, development of database for NARES library, use of Social Media, upgrade the

DSpace version from 4.x to 5.x, development of mobile application for Krishikosh, etc.

The central server of Krishikosh is being maintained by the Indian Agricultural Research Institute, New Delhi. The Administrator of Krishikosh authorises the participating libraries to upload contents to the IR. All participating libraries can log in to the Krishikosh web site using their credentials and upload contents regularly. They should upload the file and fill up all mandatory fields required to create the metadata for the document. The submission will be approved by the administrator and then a confirmation email will be sent to the library which submitted the document.

Currently, Krishikosh Institutional Repository is housing about 65,200 records of articles, books, journals, institutional publications, conference proceedings, reports, theses and dissertations, *etc.*, and the number is increasing steadily as the participating libraries are uploading documents on daily basis. The top 10 Partner Libraries which uploaded more number of documents to Krishikosh are listed in Table-2.

**Table-2: Top 10 Partner-Libraries which uploaded more documents to Krishikosh**

| Sl. No. | University | Type | Total |
|---|---|---|---|
| 1. | IARI | ICARI | 13658 |
| 2. | TANUVAS | SAU | 12707 |
| 3. | PJTSAU | SAU | 8478 |
| 4. | UAS-B | SAU | 8428 |
| 5. | IVRI-I | ICARI | 3725 |
| 6. | NDRI | ICARI | 2179 |
| 7. | MPKV | SAU | 2146 |
| 8. | ICAR | ICARI | 1946 |
| 9. | TNAU | SAU | 1895 |
| 10. | IVRI-M | ICARI | 1341 |

The following are the benefits that Krishikosh repository brings to institutions:

- Opening up outputs of NARES to a worldwide audience;
- Maximizing the visibility and impact of the outputs of NARES as a result;
- Showcasing the NARES to interested constituencies – prospective staff, prospective students and other stakeholders;
- Facilitating the development and sharing of digital teaching materials and aids, and
- Supporting student endeavours, providing access to theses and dissertations and a location for the development of e-portfolios.

## 4.2. Collections in Krishikosh

It is very important to note that Krishikosh is comprising of grey literature which are not available in the market. The following are the major documents digitized and made available under Krishikosh

- Theses
- Dissertations
- Reports
- Reprints
- Books (copyright free)
- Journal volumes
- Manuals
- Conference Proceedings
- Rare Photographs
- Convocation Addresses
- Question Papers
- Institutional Publications
- Project completion reports
- Policy documents
- Unpublished documents
- Pamphlets, etc.

Today, Krishikosh is a single point access Institutional Repository of Agricultural Knowledge for the entire NARES system which provides seamless access to the invaluable information resources online.

## 4.3. Krishikosh – A Quantitative Analysis

A recent study attempted to report a quantitative analysis of Krishikosh revealed that the KrishiKosh is housing about 65,200 records of articles, books, journals, institutional publications, conference proceedings, reports, theses and dissertations, etc., and the number is increasing steadily as the 70 participating libraries are uploading documents on daily basis. The top ten libraries have contributed 50,198 documents (76.99%) in which 5 are SAU libraries (41.95%) and 5 are ICAR Institute libraries (35.04%). IARI, New Delhi is the top contributor with 13,658 records (20.95%) followed by TANUVAS-Chennai with 12,707 records (19.49%), PJTSAU, Hyderabad with 8,478 records (13%), UAS-Bengaluru with 8,428 records (12.92%) and IVRI-Izatnagar with 3,725 records (5.71%). The remaining 15,002 documents (23.01%) are contributed by 60 partner libraries of NARES in which 33 are SAU and 27 are ICAR Institute libraries.

Among the type of documents uploaded, journals are in the top with 16,943 (25.00%) followed by Theses and Dissertations with 14,604 (22.40%), Books with 12,578 (19.29%), Articles with 3,934 (6.03%) and Reports with 2,413 (3.70%). The top five type of documents constitute 50,472 (77.41%) and the remaining documents 14,728 (22.59%) are consisting of seminar proceedings, newsletters, bulletins, book chapters, manuals, historical records, leaflets, pamphlets, photographs, *etc.*

The study also found that 28 SAUs listed in ICAR web site have not taken part in the KrishiKosh repository. Out of 70 participating libraries 8 libraries have not contributed any document in which 4 are SAU and 4 are ICAR Institutes. The study also found that 28 SAUs listed in the ICAR web site have not been taken part in the KrishiKosh which is a national level initiative.

This national level portal is possible only because of the painstaking efforts of Dr.A.K.Jain the CPI of e-Granth Project.

The team lead by Dr.A.K.Jain undertook painstaking efforts to get a unified list of documents for digitization as they wanted to ensure judicious spending on this project. The team has taken many measures to avoid duplication of records while the digitization process. A systematic review of the resource list at Institute level and national level ensured 'zero' duplication in this Institutional Repository.

## 5. Networking

The advent of computer and Internet has opened new challenges and opportunities for Indian Agriculture. Vastness and diversity of our agriculture is reflected by the fact that it caters to incredibly diverse habits and practices of our agrarian population which is spread from Kashmir to Kanyakumari. The new world economic order and globalization of markets calls for prompt and efficient infrastructure, better resource management and competitiveness of existing agricultural production system of this great farming country. It is known to everyone that Agricultural Information is vital to fulfil these dictates of time. Quick access to information at global level through electronic media thus provides the way to tackle future challenges of Indian Agriculture.

In this context, the Agricultural Research Information System (ARIS) came into being in the terminal years of the VIII plan using funds from the National Agricultural Research Project (NARP) in which Dr.A.K.Jain played a crucial role as the architect of the system.

The goal of the ARIS was to strengthen Information Management Culture using modern tools within the National Agricultural Research Information System (NARS) so that agricultural research becomes more efficient and effective. In this direction, a beginning was made by the Indian Council of Agricultural Research (ICAR), New Delhi, providing bare minimum hardware and software to 49 ICAR Institutes, 10 Project Directorates, 27 National Research Centers (NRCs), 28 State Agricultural Universities (SAUs) and 120 Zonal Research Stations (ZRSs). This enabled these institutions to get electronically connected and have E-mail facility. All the SAUs were encouraged to create Local Area Networks (LANs) in their campuses.

The major objectives of the ARIS are:

- To put information close to managers and scientists. To build the capacity to organize, store, retrieve and use the relevant information into the agricultural research infra-structure.

- To share the information over NARS and
- To improve the capacity to plan, execute, monitor and evaluate research programs.

While Dr.A.K.Jain has done computer networking through ARIS project, he played a pivotal role in human networking of library and information professionals serving across the country through the e-Granth project. He organized the e-Granth Project meeting at Delhi and invited all the library professionals of partner libraries. It was the first time many of the Agricultural Librarians came to know each other. The entire NARS library and information professionals are indebted to e-Granth Project, particularly to Dr.A.K.Jain for this real networking of Agricultural Librarians of India.

Normally, the professional societies will do the job of networking of professionals through various professional meetings, conferences, seminars, symposia, workshops, etc. There was a professional society for Agricultural Libraries which is known as "Association of Agricultural Librarians and Documentalists of India (AALDI)". But, the networking of Agricultural Librarians was done by Dr.A.K.Jain through e-Granth since the AALDI was defunct. In addition to networking the Agricultural Librarians, he has made a great opportunity to the Agricultural Librarians to revive the AALDI. He supported the revival of AALDI in all possible ways and extended his helping hands through e-Granth project to host AALDI conferences. Thus, the real networking of Agricultural Librarians was done by Dr.A.K.Jain under the umbrella of e-Granth project.

## 6. Conclusion

Many initiatives had been taken under ICAR funded projects to strengthen the library and information services of the NARES system. However, the e-Granth Project, which was led by Dr.Arun Kumar Jain has made tremendous changes in the functioning of the libraries of NARES. In fact, the real digital library services were started in NARES only after the launch of e-Granth project and CeRA for which the entire NARES community is always indebted to the team lead by Dr.Arun Kumar Jain, the man who brought effective and efficient ICT applications in the NARES system including the libraries.

## Acknowledgement

I would like to record here the unstinted support extended by Dr.C.Devakumar, Former Assistant Director General (EP & D), Indian Council of Agricultural Research, New Delhi to the entire e-Granth Team to achieve the goals of the project. The e-Granth team always indebted to Dr.H.Chandrasekaran, PI of CeRA and Co-PI of e-Granth project, Dr.P.S.Pandey, National Director Component-1 of NAIP who took a lot of efforts to support the Library and Information Science professionals of NARES system many ways.

## References

Gutam, Sridhar. "KrishiKosh-An Institutional Repository under National Agricultural Research System." (2013). Available at https://works.bepress.com/sridhar_gutam/37/

Jain, A. K. *Strengthening of Digital Library and Information Management under NARES (eGranth).* IARI, New Delhi, 2014.

Jain, Arun K., and K. Veeranjaneyulu. "Repository Of Indian National Agricultural Research & Education System (NARES): Open Access To Institutional Knowledge." (2013).

Rathinasabapathy, G., K.Veeranjaneyulu and Amarenrakumar (2016). Krishikosh: Institutional Repository of National Agricultural Research and Education System (NARES) of India: An Analytical Study. In: Conference: International Conference of Digital Libraries (ICDL) 2016: Smart Future - Knowledge Trends that will Change the World, At New Delhi, India, Volume: 1

Roy, Bijan Kumar, Subal Chandra Biswas, and Parthasarathi Mukhopadhyay. "AgriCat: An One-stop Shop for OAI-based Open Access Agricultural Repositories." *AGRÁRINFORMATIKA/ JOURNAL OF AGRICULTURAL INFORMATICS* 7.1 (2016): 103–115.

Varshney, Himanshu, and Aruna Kumar. "Knowledge Management at ICAR (Indian Council of Agricultural Research)." (2012).

Veeranjaneyulu, K. "KrishiKosh: an institutional repository of National Agricultural Research System in India." *Library Management* 35.4/5 (2014): 345–354

# 8

# Mobile Apps in Agriculture: An Overview

**Dr.D.Chandran[1] and Dr. Kumar Kutty[2]**

[1] Former Professor, Dept. of Library & Information Science Sri Venkateswara University,Tirupati-517501, Andra Pradesh

[2] Librarian-in-charge, College of Veterinary Science Library Sri Venkateswara Veterinary University, Proddatur, Andra Pradesh

**ABSTRACT**

*Information and communication Technology incorporate collections from radio to satellite, to mobile phones or electronic money transfers. The increase in their affordability, ease of access, and flexibility has resulted in their use even within rural homesteads relying on agriculture. Extensive implementation of android mobile applications has gained conventional fame. However, the prospective privacy and security risks associated with using mobile apps are quite high, as smart phones become progressively incorporated with our lives, being able to access our email, social networking accounts, financial information, individual photo & even our cars and homes. This paper addresses mobile applications for agriculture that are designed to convey information either visually or audibly make them useful to farmers with limited formal education or exposure to technology.*

***Keywords:*** *Mobile, Apps, Agriculture, ICT*

# 1. Introduction

Mobile apps are software designed to take advantage of mobile technology and enveloped for technology besides mobile phones. However, mobile phones have many key advantages: affordability, possession, voice interactions, and instant and convenient service delivery. As a result, there has been a global explosion in the number of mobile apps, facilitated by the rapid evolution of mobile networks and by the rising functions and declining prices of mobile handsets.

It estimated that, by 2016, the number of smart phone users would be more than 2 billion people worldwide. Internet and Mobile Association of India (IAMAI) and consultancy firm KPMG reported that the number of mobile Internet users in India projected to cross 503 million marks by 2017 from 427 million users at present. Many industries have adopted smart phones to facilitate their work, such as health care and education.

This paper overviews the usage of smart phone applications in an important sector, agriculture. Mobile based applications in agriculture are an emerging technology, focusing on the enhancement of agricultural and rural development in India. The advancements in mobile applications can be utilized for providing precise, sensible, pertinent information and services to the farmers, thus facilitating an environment for more remunerative agriculture.

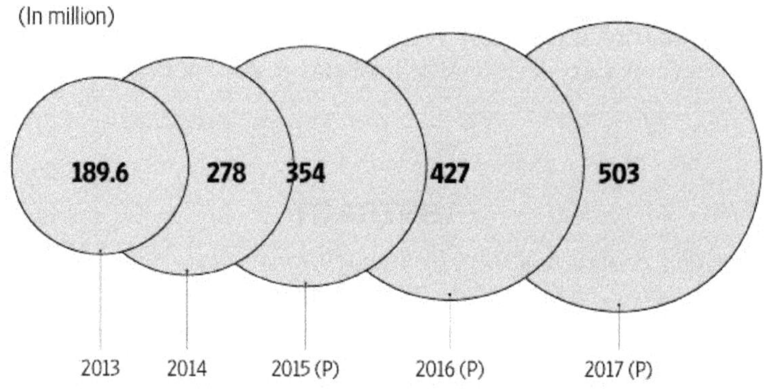


**Figure 1: Growth of Mobile Internet Users in India**

Traditional agriculture usually done within a family or a village and accumulative farming expertise and knowledge passed down to their future generations. Human was the main observer of field background and the solution contributor as problems arise. However, this traditional method is no longer suitable, because farming outputs rely fundamentally on the natural surrounding atmosphere (e.g. weather and water) and global warming concerns (causing frequent droughts and floods) & crop disease outbreaks are troublesome of farming productivity.

For decades, farmers depended on the radio broadcasting and television to find out what was going on in the product markets. At present, technology is universal, using mobile applications[9] farmers are now able to check commodity prices, make out how much fertilizer or pesticide should applied at a specific point in a larger field, find out the temperature or moisture in the grain bin need to be adjusted.

## 2. Mobile APPs: Its importance for Agriculturalists

Agricultural sector is among the most imperative production sectors in the world since it is the main food supplier. However, agricultural sector is one of the business sectors that left aside in terms of the application of new technologies. Stakeholders involved in the agricultural sector are progressively use ICT know-how and sometimes with remarkable proportions. Most mobile apps for agriculture and rural development spotlight on enlightening agriculture supply chain integration and have a wide range of functions, for instance providing market information, growing access to extension services, and facilitating market links. Users are also assorted, including farmers, produce buyers, cooperatives, input suppliers, content providers, & other stakeholders who demand useful, affordable services. These mobile-based applications could provide momentous economic and social benefits—among them, create jobs, adding value, reducing product losses, and making developing countries more globally competitive. Drawing on simple, available technologies such as SMS, service providers can offer mobile banking, and other transaction services (selling inputs, for example), and information services (market price alerts). Table 1 shows various objectives of mobile applications in agriculture.

**Table 1: Various purpose of Mobile Apps in Agriculture**

| S.No | Purpose | Method |
|---|---|---|
| 1 | Education and awareness | Information provided via mobile phones to farmers and extension representative about good practices, improved crop varieties & pest or disease management. |
| 2 | Commodity prices and market Information | Prices in regional markets to inform decision making throughout the entire agricultural process. |
| 3 | Data Collection | Applications that collect data from large geographic regions. |
| 4 | Pest and disease outbreak warning and tracking | Send and receive data on outbreaks |

## 3. Mobile Apps in Agriculture (selective)

In a developing country like India, technology[14] shows its trace in the least of villages. Approximately all over now, there is a mobile tower in sight and people seen using mobile phones. Farming and agriculture has never been more competitive. Facing the challenges of a changing climate, growing costs, and global difficulty on grain and feed, farmers need to make the most out of existing resources to maximize profits. Mobile apps[15] can help farmers and agriculture businesses stays connected in real-time to workers in the field, and develop competence throughout the entire operation. Table 2 presents information on mobile apps for farmers[16-33] to increase efficiency and generate higher profits.

**Table 2: List of Mobile Apps for Agriculture**

| S.No | Category | Apps | Feature |
|---|---|---|---|
| 1 | Weather. | Weather Bug | Latest temperature, humidity levels, visibility, severe weather alerts or forecast |
| 2 | Prices. | Cash Grain Bids<br>Cash Grain Bids Price Finder | Latest crop or livestock prices |
| 3 | Pests | MyTraps<br>mytraps | Monitor levels and movements of insects throughout your fields to help better target insecticide applications. |
| 4 | Soil | SoilWeb | Works with the phone's GPS system to help identify soil properties and assist with soil sampling needs |
| 5 | Farm Management | Virtual Farm Manager,<br>VirtualFarm MANAGER | provides a number of tools to store, view and log information about one's farm |
| 6 | News Reader. | USDA News Reader.<br>USDA Dept of Agriculture News Reader | Create your own news feeds, learn about recovery plans and programs, and easily navigate the massive USDA website in a mobile-friendly environment. |
| 7 | Livestock producers | Livestock Manager<br>Livestock Manager | Allows users to track various information about their animals, as well as carrying information, medicine administration and more. |

**Table 2: (Continued)**

| S.No | Category | Apps | Feature |
|---|---|---|---|
| 8 | Material safety data | Agrian Mobile | Browse usage rates, pre-harvest and re-entry intervals, worker safety information and more. |
| 9 | Better Record Keeping | Mix Tank 2.0 | Calculate the proper mixing order of tank mix. After determining the tank-mix parameters, you be able to make field notes on volume per acre, nozzle type, etc., |
| 10 | Location based F2S (Farm to Shop) Trading Platform for Agricultural Product. | Mandi Trades | One can get information about commodity prices directly provided by the Government of India, A buy/sell platform aims to solve the problems of marketing, and farmers can avoid the daily struggle to sell their produce. |
| 11 | Host of services | Rainbow | 1. Live market prices from various markets in India.<br>2. Daily dam water level.<br>3. Location based weather updates.<br>4. Localized latest agro news.<br>5. Call/SMS option for connecting to seed, fertilizer and pesticide dealers in a single tap. |
| 12 | Live price updates of commodities from various markets across India. | Market Watch | Provides the wholesale price information about commodities as per the datasets |
| 13 | Prices of various commodities in India | Gramseva: Kisan | Check commodity prices without the Internet.<br>Generates a graph based on the Historical data for analysis. |

(Continued)

**Table 2: (Continued)**

| S.No | Category | Apps | Feature |
|---|---|---|---|
| 14 | Government staff advisories | mKisan | Free of cost and do not have any royalty or Intellectual Property Right (IPR) issue. It enables farmers and all other stakeholders to obtain advisories and information being through experts & govt. officials at different levels through mkisan portal without registering on the portal. |
| 15 | Update of Market Prices | Agrimarket  | This app automatically captures the location of the farmers using mobile GPS & fetches the bazaar cost of crops in markets which fall within the range of 50km. There is another option to get price of any market and any crop in case farmers do not want to use GPS feature |
| 16 | Information for Rainfall | Sirrus  | soil sample, scout your crops, check weather condition & graph rainfall guess per field |
| 17 | Local and State Reports | Great Lakes Early Detection Network (update)  | These reports are uploaded to EDDMaps and e-mailed directly to local and state verifiers for review. |
| 18 | Weather and Forecast details | Agrivi 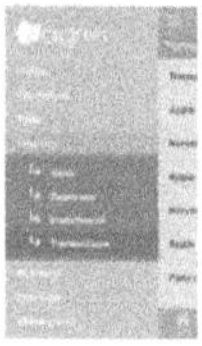 | Monitoring and tracking all farm activities & effort usage, progress on sales and expense tracking ensures taking control more farm finances, inventory management with small stock alarms eliminate delays in production caused by require of say and weather supervise with detailed 7-day weather predict & 3-year weather history for each field and smart disease risk detection alarms |
| 19 | Agriculture machines | TractorPal  | This app keeps inventory & maintenance report for all your individual agriculture machines and attachments, with cars, trucks & all brands. TractorPal enables your all machinery and automobiles include tractors, pickups, lawn mowers, cars, combines, sprayers, loaders, skid-loaders, backhoes, attachments, and more |

**Table 2: (Continued)**

| S.No | Category | Apps | Feature |
|---|---|---|---|
| 20 | Tire informations | Tire Advisor | Currently you can contact all of Bridgestone's commercial tire products in one easily searchable app. Included are truck and bus tires |
| 21 | Agricultural Machinery information | MachineryGuide | It can be used for any agricultural activity which is through by tractor or other agricultural machinery, including fertilization, manure application and spraying. It even can be use for ground measurements as well |

# 4. Global Perspective

The number of mobile applications for Agriculture & Rural Development (m-ARD) identified by Country Based on the typology, India[34] and Kenya have the most m-ARD apps Figure 2) and the broadest spread of subsector and segment activities.

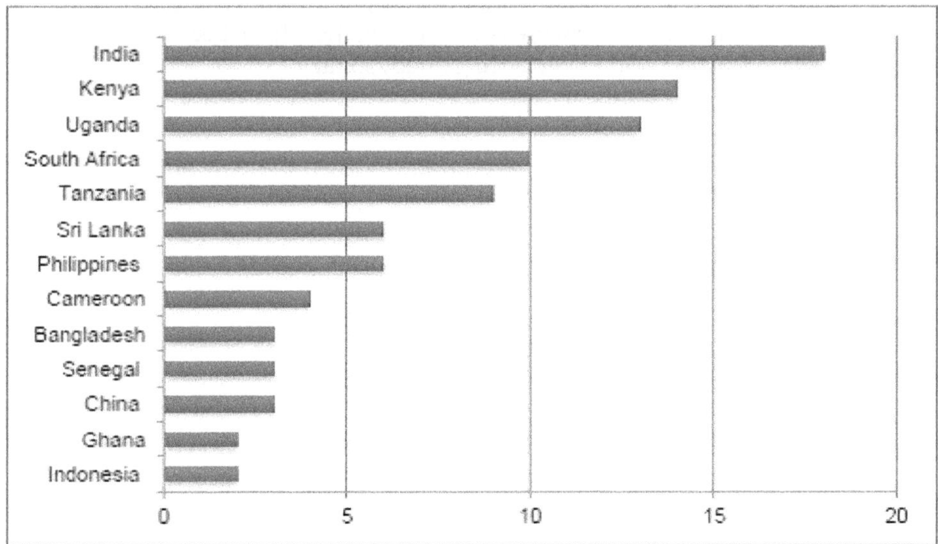


**Figure 2: Country wise distribution of Mobile Apps for Agriculture & Rural Development**

## 5. Indian Scenario

In India today, more people have access to mobiles than to running water, "These latest apps present an opportunity like no other to revolutionize life for farmers there." Farmers make up about 24.80% of India's total workforce, with over 118.9 million farmers spread across the length & breadth of the country. Agriculture is the biggest trade supplier in India contributing a significant percentage toward its GDP. It has witnessed green, yellow and blue revolution & currently it is a proud participant of the mobile revolution too. "Government spends vast quantity in extending harvest insurance to farmers. Due to administrative and technical cause, greatly of the information associated to crop insurance has not able to reach farmers in time to take advantage of the existing schemes.

## 6. Conclusion

The mobile phenomenon is especially important for developing countries since that is anywhere it is increasing fastest, and in the next few years, nearly all new mobile customers will come from developing countries since dissemination has reached diffusion levels in developing countries. Taken a massive leap beyond the costly, bulky, energy-consuming equipment once available to the very few to store and analyze agricultural and scientific data. With the booming mobile applications, ICT has found a foothold even in poor smallholder farms and in their activities. However, not all the ICT initiatives are uniform with disparities between regions in the level and excellence of telecommunications, information and the endeavour of individuals, public and private organizations, and differentiated nature of the demand of the farmers in different areas. As a result, there have been many successes, failures, lessons learned and experience gained, so far.

## References

Cheung, W.S. and K. F. Hew, "A review of research methodologies used in studies on mobile handheld devices in K-12 and higher education settings," Australasian Journal of Educational Technology, vol. 25, no. 2, pp. 153–183, 2009.

Christine Zhenwei Qiang, Siou Chew Kuek, Andrew Dymond and Steve Esselaar. Mobile Applications for Agriculture and Rural Development. ICT Sector Unit World Bank December 2011 Available at: http://siteresources.worldbank.org/INFORMATIONANDCOMMUNICATIONAND TECHNOLOGIES/Resources/MobileApplications_for_ARD.pdf

Doering C. "Farmers growing comfortable with mobile devices." March 3, 2013. USA Today website www.usatoday.com. Available at http://www.usatoday.com/story/ news/nation/ 2013/03/03/ farming-technology-ipad-apps/1959139/.

Habib, M.A., M. S. Mohktar, S. B. Kamaruzzaman, K. S. Lim, T. M. Pin, and F. Ibrahim, "Smartphone-based solutions for fall detection and prevention: challenges and open issues," Sensors, vol. 14, no. 4, pp. 7181–7208, 2014.

Hopkins M."10 Best Mobile Agriculture Apps For 2012." March 15, 2012, (UPDATED: March 25, 2013). Available at: http://aged.illinois.edu/sites/aged.illinois.edu/files/resources/Apps-for-Ag-Revised.pdf

Milrad, M and D. Spikol, "Anytime, anywhere learning supported by smart phones: experiences and results from the musis project," Educational Technology and Society, vol. 10, no. 4, pp. 62–70, 2007.

Mosa, S.M. I. Yoo, and L. Sheets, "A systematic review of healthcare applications for smartphones," BMC Medical Informatics & Decision Making, vol. 12, no. 1, article 67, 2012.

Potter B. "10 Handy Farm Apps." October 16, 2012. AgWeb website www.agweb.com. Available at http://www.agweb.com/article/10_handy_farm_apps/

Sasson, "Food security for Africa: an urgent global challenge," Agriculture & Food Security, vol. 1, no. 2, pp. 1–16, 2012.

Sjostrom L. "Dairy farming apps could pay back." September 5, 2012. Hoard's Dairyman Available at http://www.hoards.com/blog_dairy-apps.

Sotiris Karetsos , Constantina Costopoulou , Alexander Sideridis. Developing a smart phone app for m-government in agriculture Journal of Agricultural Informatics . 2014 Vol. 5, No. 1:1–8

# 9

# Transformation of Agricultural and other Libraries in Digital Era

**Arundhati Kaushik**

University Library
G.B. Pant University of Agriculture and Technology
Pantnagar – 263145, Uttarakhand

## 1. Agricultural Libraries in India

Agricultural libraries in India are in transition mode from repositories to open access, vanishing gap between user needs and library services, opting ICT technologies for dissemination of information. It is changing dramatically by adopting new Information Communication Technologies in all housekeeping activities of print to e-environment like printed library card catalogue to Online Public Access Catalogue, online accessibility for 24/7, availability of e-databases, access to online databases, e-journals, online journals. These libraries are being act as: Gateway to information; Training centre; Publication centre; Learning centre and Disseminator of information literacy.

One primary role of Agricultural Librarians is to provide leadership and expertise in the design, development, and ethical management of knowledge-based information systems in order to meet the information needs and obligations of the patron or academic institution. At present many virtual libraries have been created under ICAR system and library professionals enriching their management skills to play leadership role in the digital future for organizing managing and disseminating e-literacy to users.

## 2. Digital transformation

The digital revolution era started in 1980s and is also called the Third Industrial Revolution refers to the advancement of technology from analog electronic and mechanical devices to the digital and online information communication technology available today. It also marks the digital Information era which transformed the libraries roles in the society.

## 3. Open access transformation

Open access (OA) movement continues to transform the global research communication and dissemination system. OA research materials have come to occupy an increasing share of scholarly research dissemination across diverse publishers, geographical regions and scientific disciplines. Open accessible electronic institutional repositories that may include already published articles (post-prints), pre-published articles (pre-prints), theses, manuals, teaching materials or other documents that the authors or their institutions wish to make publicly available without financial or other access barriers.

With the fast evolving globalization of agriculture and importance of information management its communication and dissemination has put the Indian Council of Agricultural Research, New Delhi to think to take leadership and to lay down dynamic policies to develop information communication development activities in India. The key features of these policies are as follows:

- Strengthening information, communication and dissemination systems;
- Enhancement of public awareness capacity and improved knowledge sharing;
- E-learning, distance learning and developing training materials to prepare users with newly required skills and knowledge;
- Assessing and adapting the current learning and capacity building and human resource development initiatives

## 4. Proactive role of Dr. A.K. Jain, Professor, CPI, e-Granth

The role of the Leader is from a passive intermediary role responsible for guiding to appropriate information to analyzing and repackaging information, content information management and digital repository management systems. **Dr. A.K. Jain** has taken dynamic initiatives and given opportunities to Library Professionals for:

- Developing digital libraries as natural complements to digital learning environment to support educators with respect to the selection of adequate resources for a given source;

- Managing, organizing, indexing and integrating digital information from the resources to the users of the library;
- Teaching information literacy to educate future knowledge workers, in traditional ways or via internet-based instruction modules;
- Providing a learning centre to serve as a physical learning environment suitable for more active learning styles (Pg 143-144)
- To become information generators, gatherers, recorders, processors, organizers, disseminators, retrievers, preservers, measurers and compilers in the digital environment.

## 5. As a leader he faced challenges on:

**Digital divide:** Innovations in ICT are dividing the world in different divisions with reference to infrastructural facilities. ICAR, New Delhi with all its efforts have created basic infrastructure for free flow of information and converting agricultural systems as vibrant knowledge super-hub.

**Information Literacy:** Handling and exploration of the information is very important, therefore, it is the duty of a Librarian to teach and train the user to retrieve the relevant information of their respective subject areas. Information literacy is the concept dealing with management, retrieval and dissemination of information which is available online or through internet only.

**Quality of Contents:** Repositories to be uploaded should be of quality contents. These should be reviewed by the subject specialist.

**Content creation, management and dissemination:** Creation of scholarly contents its accumulation and providing interactive interface to the end user. Various content management systems with flexibility and interactive search options are required.

**Specifications and Standards:** Set of rules and specifications are required for longer usability, re-usability, accessibility and availability of all scholarly content. For this a set of rules and specification are required.

**Copyright:** To make scholarly communication available in the public domain there should be well-defined policy regarding copyright issues.

## 6. Open Access Movement and Academic Institutional Repositories

The core essence of open access movement is to "make research articles in all academic fields freely available on the internet". The open access journal models and institutional repositories are being experimented by the educational institutions notably the universities.

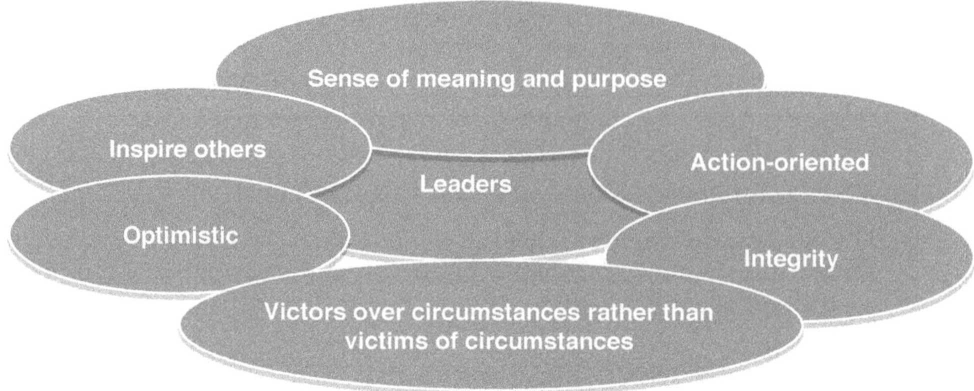


**Fig. 1 Leadership qualities**

## 7. Union catalogue and digital libraries

Union catalogue and digital libraries are the new paradigms of modern library. A union catalogue is a combined library catalogue describing the collections of a number of libraries. They have been created in a range of media, including book format, microform, cards and more recently, networked electronic databases. Many agricultural libraries in India have taken up the challenges of modernization to develop the Machine Readable Catalogue (MARC) and machine readable full text documents to provide greater accessibility to full text databases of different nature, changing the focus of simply providing library services to fulfilling the long pending focus of Right to Information.

Union catalogue are useful to librarians as they assist in locating and requesting materials from other libraries through interlibrary loan service. With the help of Online Computer Library Centre (OCLC), USA our libraries are in position to be discoverable by the network of global library network and sharing of online resources more effectively.

### 7.1. WorldCat

WorlCat is the world's largest network of library content and services. WorldCat libraries are dedicated to providing access to their resources on the web, where most people start their search for information. It lets you search the collections of libraries in your community and thousand more around the world.

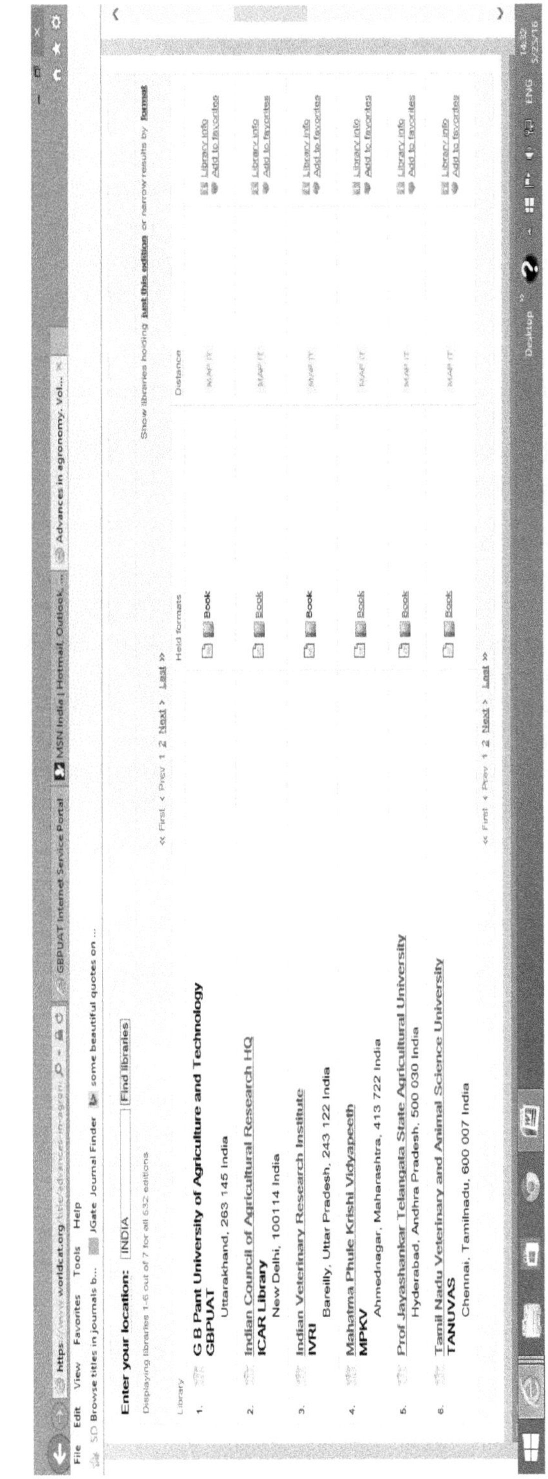

Enter your location: INDIA  Find libraries

Displaying libraries 1-6 out of 7 for all 632 editions

| Library | Held formats | Distance |
|---|---|---|
| 1. G B Pant University of Agriculture and Technology GBPUAT Uttarakhand, 263 145 India | Book | MAP IT |
| 2. Indian Council of Agricultural Research HQ ICAR Library New Delhi, 100114 India | Book | MAP IT |
| 3. Indian Veterinary Research Institute IVRI Bareilly, Uttar Pradesh, 243 122 India | Book | MAP IT |
| 4. Mahatma Phule Krishi Vidyapeeth MPKV Ahmednagar, Maharashtra, 413 722 India | Book | MAP IT |
| 5. Prof Jayashankar Telangata State Agricultural University Hyderabad, Andhra Pradesh, 500 030 India | Book | MAP IT |
| 6. Tamil Nadu Veterinary and Animal Science University TANUVAS Chennai, Tamilnadu, 600 007 India | Book | MAP IT |

**Fig. 2 Agricultural Libraries of India on WorldCat**

### 7.2. AgriCat

AGRICAT Libraries is dedicated to offering its access to the widest possible range of resources. One can now search its collection and the collections of many other libraries worldwide using WorldCat, the world's largest network of library-based content and services. Use simple keywords and more important try to use words from a title; the name of the author or words that describe the subject matter.

AgriCat 2.0 is a Union Catalog of Library Resources of Indian National Agricultural Research & Education System (NARES). It intends to provide open access to all the catalog of the holdings of NARES libraries so that the sharing of library resources becomes possible. One can easily locate resource availability and the complete details and previews in some cases. At present 38 libraries is part of this acitivity. The NAIP which aims to facilitate accelerated and sustainable transformation of Indian Agriculture through innovation has supported this development though eGranth sub-project.

## 8. Other initiatives in India

### 8.1. IndCat

Online Union Catalogue of Indian Universities is unified Online Library Catalogues of books, theses and journals available in major university libraries in India. The union database contains bibliographic description, location and holdings information for books, journals and theses in all subject areas available in more than 178 university libraries across the country. it consists three components available in open access to users and librarians.

- **Books:** Over thirteen millions bibliographical records of books from 178 university libraries.
- **Theses:** Doctoral theses submitted to various Indian universities till date.
- **Serials:** Currently subscribed journals by the universities and holdings information on serials available in various university libraries.

It is a major source of bibliographic information that can be used for inter-library loan, collections development as well as for copy cataloguing and retro-conversion of bibliographic records. The records can be downloaded and can easily merged in to SOUL software or any another software which adheres to MARC 21.

### 8.2. DELNET (Developing Library Network)

DELNET was started at the India International Centre Library in January 1988 and was registered as a society in 1992. It was initially supported by the National

Information System for Science and Technology (NISSAT), Department of Scientific and Industrial Research, Government of India. It was subsequently supported by the National Informatics Centre, Department of Information Technology, Ministry of Communications and Information Technology, Government of India and the Ministry of Culture, Government of India

DELNET has been established with the prime objective of promoting resource sharing among the libraries through the development of a network of libraries. It aims to collect, store, and disseminate information besides offering computerised services to users, to coordinate efforts for suitable collection development and also to reduce unnecessary duplication wherever possible.

## 8.3. Union Catalogue of Biomedical Serials

A database of the serials holdings of major medical libraries in the country has been compiled by Indian Medlars Centre (IMC). This serves as an important information tool for locating journals in 188 libraries in India. The database is regularly updated and can be accessed by users free of cost. The database can be accessed by journal title, state and library.

## 8.4. Union Catalogue of Serials (National Union Catalogue of Scientific Serials)

National Union Catalogue of Scientific Serials in India (NUCSSI) is the first indigenous database that serves as an ideal tool to access journal holdings information. Journals are the main source of Science & Technology information. NUCSSI is a data repository of a large number of unique journal titles and library holdings belonging to major universities, R&D units of industries, higher institutes like IISc, IITs and professional institutes spread all over the country.

NUCSSI provides:

- Information regarding the availability of journal title in libraries, moreover integration of database with e-mail service enables routing of library/user request information.
- The regular update of the database is enhanced with the online access granted to the participating libraries via internet, so that journal seekers can get the updated information free of cost.
- User friendly interface and powerful search enables easy and improved access to locate a particular journal and its availability in various libraries.

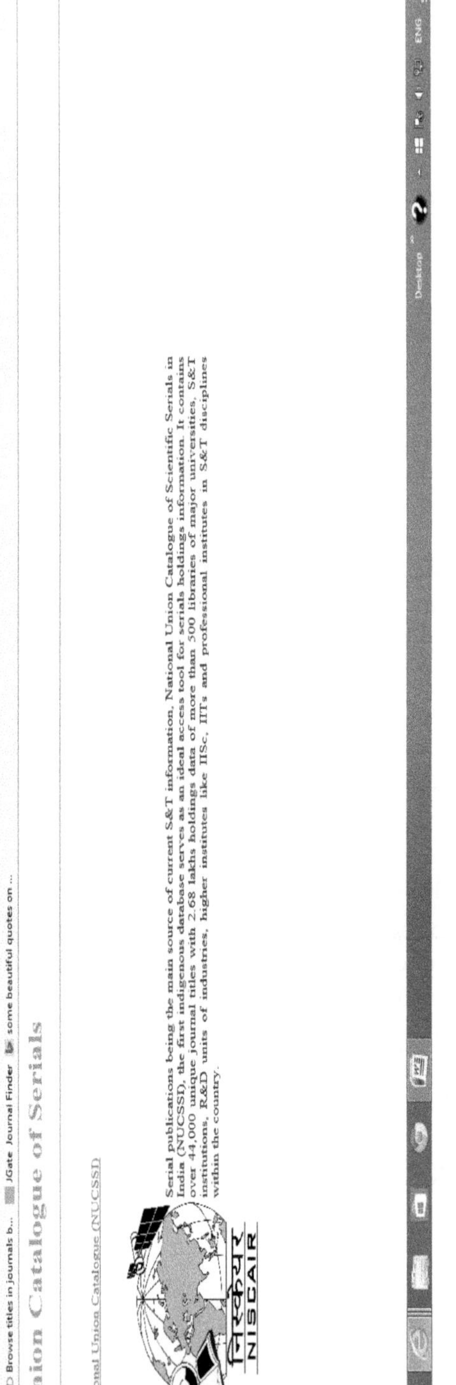
Union Catalogue of Serials
National Union Catalogue (NUCSSI)
निस्केयर
NISCAIR
Serial publications being the main source of current S&T information, National Union Catalogue of Scientific Serials in India (NUCSSI), the first indigenous database serves as an ideal access tool for serials holdings information. It contains over 44,000 unique journal titles with 2.68 lakhs holdings data of more than 500 libraries of major universities, S&T institutions, R&D units of industries, higher institutes like IISc, IITs and professional institutes in S&T disciplines within the country.

**Fig. 3 NUCSSI**

## 8.5. The Indian National Bibliography

**The Indian National Bibliography** has been conceived as an authoritative bibliographical record of current Indian publications in Assamese, Bengali, English, Gujarati, Hindi, Kannada, Malayalam, Marathi, Oriya, Punjabi, Sanskrit, Tamil, Telugu and Urdu languages, received in the National Library, Kolkata under the Delivery of Books and Newspapers, (Public Libraries) Act, 1954 (Act No. 27 of 1954 as amended by Act No. 99 of 1956). Maps, Musical scores, Periodicals and Newspapers (except the first issue of a new periodical and the first issue of a periodical under a new title), Keys and Guides to Textbooks and Ephemeral and other such materials

The main entries are in Roman Script and the collations and annotations, if any, are in English. The classified portion follows the Dewey Decimal Scheme of Classification but the numbers from the Colon Classification scheme are assigned to each entry at the bottom right hand to facilitate the use of the Bibliography and libraries, arranged according to the Colon Schemes of classification. INB records from since inception (1958- ) has been converted into electronic data.

## 8.6. National Library Automation & Resource Sharing Network (N-LARN)

The aim of the project is to develop a national union database (Union Catalogue) of resources in libraries of higher educational institutions with focus on library resources in colleges and training of library professionals in library automation. The entire project is funded and supported by the Ministry of Human Resources Development, Government of India under its National Mission for Education through ICT (NMEICT) programme. Under the project series of workshops are being conducted in various parts of the country to train LIS professionals to use appropriate library automation software (KOHA) and relevant standards such as MARC 21. The workshops provide hands-on-training to participants to enable them to implement the software and standards in their respective institutions. The participating institutions are expected to provide bibliographic data for resources available in their libraries to populate the union database. The bibliographic data received from participating institutions will be edited, updated and made available on a centralized server as part of the union database. Libraries participating in the project will also be able to download standard records from the union database to hasten the process of automating their libraries. The project also supports libraries without adequate infrastructure to use the centralized facility for automating their libraries.

## 8.7. National Library and Information Services Infrastructure for Scholarly Content (N-LIST)

The Project entitled "National Library and Information Services Infrastructure for Scholarly Content (N-LIST)", being jointly executed by the UGC-INFONET Digital Library Consortium, INFLIBNET Centre and the INDEST-AICTE Consortium, IIT Delhi provides for i) cross-subscription to e-resources subscribed by the two Consortia, i.e. subscription to INDEST-AICTE resources for universities and UGC-INFONET resources for technical institutions; and ii) access to selected e-resources to colleges. The N-LIST project provides access to e-resources to students, researchers and faculty from colleges and other beneficiary institutions through server(s) installed at the INFLIBNET Centre. The authorized users from colleges can now access e-resources and download articles required by them directly from the publisher's website once they are duly authenticated as authorized users through servers deployed at the INFLIBNET Centre.

**N-LIST: Four Components:** The project has four distinct components:

i. To subscribe and provide access to selected UGC-INFONET e-resources to technical institutions (IITs, IISc, IISERs and NITs) and monitor its usage;

ii. (The INDEST and UGC-INFONET)

iii. To subscribe and provide access to selected INDEST e-resources to selected universities and monitor its usage;

iv. (The INDEST and UGC-INFONET)

v. To subscribe and provide access to selected e-resources to 6,000 Govt. / Govt.-aided colleges and monitor its usage;

vi. (The INFLIBNET Centre, Ahmedabad)

vii. To act as a monitoring agency for colleges and evaluate, promote, impart training and monitor all activities involved in the process of providing effective and efficient access to e-resources to colleges.

(The INFLIBNET Centre, Ahmedabad)

The INFLIBNET Centre is also responsible for developing and deploying appropriate software tools and techniques for authenticating authorized users.

## 8.8. ShodhGangotri

The word **"Shodh"** originates from Sanskrit and stands for "research and discovery". **"Gangotri"** is one of the largest glaciers in the Himalayas and source of origination of Ganges, the holiest, longest and largest of rivers in India. The Ganges is the symbol of age-long culture, civilization, ever-aging, ever-flowing, ever-loving and loved by its people.

Under the initiative called **"ShodhGangotri"**, research scholars / research supervisors in universities are requested to deposit electronic version of approved synopsis submitted by research scholars to the universities for registering themselves for the Ph.D programme. The repository on one hand, would reveal the trends and directions of research being conducted in Indian universities, on the other hand it would avoid duplication of research. Synopsis in "ShodhGangotri" would later be mapped to full-text theses in **"ShodhGanga"**. As such, once the full-text thesis is submitted for a synopsis, a link to the full-text theses would be provided from ShodhGangotri to **"ShodhGanga"**

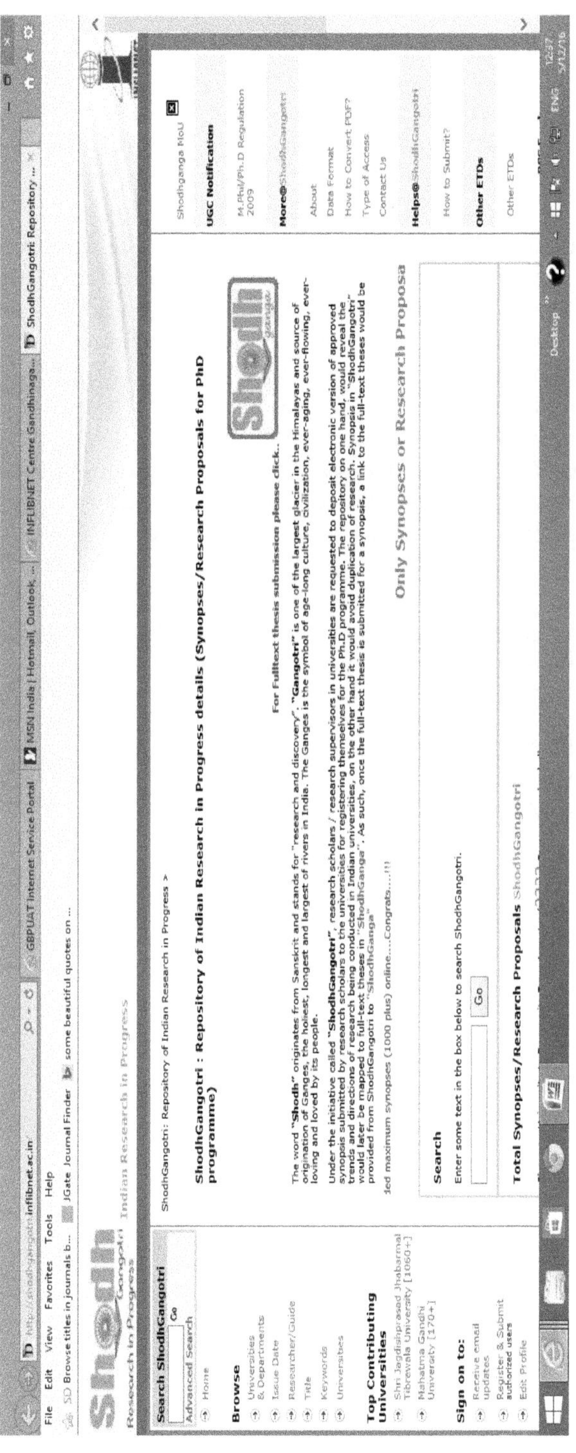


Fig. 4: Indian Research in progress

## 8.9. CSIR Virtual Union Catalogue

Data of 31 KRCs of CSIR laboratories have been migrated to koha successfully OPAC of 31 KRCs of CSIR laboratories have been published on Internet Data of 29 KRCs of CSIR laboratories have been harvested to CSIR Virtual Union Catalogue successfully Currently, CSIR Virtual Union Catalogue contains around 10 lakh records.

# 9. Conclusion

Many Agricultural and other libraries in India have already started in the direction to incorporate advanced digital technologies to meet the needs of their users. These libraries are more focused on digitization, digital collections, preservation and data storage and retrieval. According to the words of Clifford Lynch "Digital technologies have opened the door to a host of new possibilities for sharing knowledge and generated entirely new forms of content that must be made broadly available. This shift demands that universities take on a much more active role in ensuring dissemination of the knowledge produced by their institutions – both now and in the future". Now it is the responsibility of Information Professionals in India to examine how they can adopt new technologies more efficiently.

# References:

Delhi University Library System, University of Delhi. 2010. Globalizing academic libraries: Vision 2020. Delhi: Mittal Publications. 2 Volumes. 711p.

http://2cqr.in:3000

http://egranthalaya.nic.in

http://inboline.nic.in

http://indcat.inflibnet.ac.in

http://indmed.nic.in/trng/uniondata.pdf

http://knowgate.niscair.res.in/catalogue

http://n-larn.ac.in

http://ntiindia.kar.nic.in

http://nucssi.niscair.res.in

http://sauc.uchicago.edu

http://shodhgangotri.inflibnet.ac.in

http://www.techopedia.com

nlist.inflibnet.ac.in

www.delnet.nic.in

www.egranth.ac.in/agricat.html; http://agricat.egranth.ac.in

www.en.wikipedia.o

www.icast.org.in

www.iitgn.ac.in

www.researchgate.net

www.rri.res.in/library.html

www.worldcat.org

# 10

# Library Consortia in India and its Impact

**Dr. Raj Kumar Singh**
Librarian
College of Horticulture and Forestry, Central Agricultural University,
Pasighat, Arunachal Pradesh, India
email: singhraj1963@gmail.com

## 1. Introduction

A library consortium is a group of libraries whose partner to coordinate the activities, share resources, and combine expertise. Library consortia offer significant advantages to increasingly strapped libraries. The sharing of electronic information resources and collaboration on shared goals often enable libraries to deliver higher quality services than they would be able to deliver on their own. The term consortia may be defined as "A cooperative arrangement of purchasing electronic recourses among a group of institutions, which will provide collective purchasing power and enable them to avail best possible bargaining facility to ensure highest discount price for electronic journals". Library Consortia is an association or a partnership has long being a principle of librarianship. It refers to cooperation, coordination and collaboration between libraries for the purpose of sharing information resources and services. Rajgoli, Birdie & Karisiddappa, (2005) stated that a consortium is a co-operative arrangement among groups or institutions or an association or society. Consortia are commonly formed to increase the purchasing capacity of the collaborating institutions, to expand the resource availability and to offer automated services. In other words, it is described as a group of organizations whose purpose is to collectively facilitate and support the work of a service program in ways that ad material and human resources beyond those available to each organization individual. Pathak and Deshpande, (2004) discuss the importance of consortia and their impact on society especially in developing countries like India. The present scenario of consortia among university libraries, special libraries and different technical institute libraries in India is elaborated. Francis (2005) depicted that the new model of library consortia in which all academic intuitions and government research organizations could participate. The formation of consortia under the direction and full support by the government of India is stressed.

## Benefits to Consortia users:

- Space and time –invariant
- Helps save time for users
- Provides value addition such as search ability, alerting services, links from one article to another, forward links like citing articles etc.
- Accelerates publications.
- Articles cannot be mutilated, stolen or lost.

## Benefits to Librarians: -

- Speedup resources delivery.
- Reduced shelving, binding, maintenance, clamming, etc.
- Cost saving.
- Time saving at work.
- Improved services
- Public relation opportunities
- Greater visibility of products.

## 2. Library Consortia in India: -

There are some consortia functioning in India as per discipline wise in the educational institution and universities which can be seen here as under.

| S.N. | Name of Consortia | Members | No. of Resources | Group of Publishers | URL (User Resource Locator) |
|---|---|---|---|---|---|
| 1. | UGC INFONET2.0 (INFLIBNET) | 422 | 7500 | 26 | http://www.inflibnet.ac.in/infonet |
| 2. | CeRA | 134 | 3000 | 15 | http://www.cera.jccc.in |
| 3. | ERMED | 98 | 1812 | 10 | http://www.nmlermed.in |
| 4. | INDEST | 82 | 6500 | 17 | http://www.paniit.iitd.ac.in/indest |
| 5. | NKRC | 68 | 4500 | 23 | http://ejournal.niscair.res.in/index.php |
| 6 | DeLCON DBT e-Library Consortium | 33 | 917 | 19 | http://delcon.gov.in |
| 7. | FORSA | 11 | 25 | 10 | http://www.ncra.tifr.res.in/library/forsaweb/index.htm |

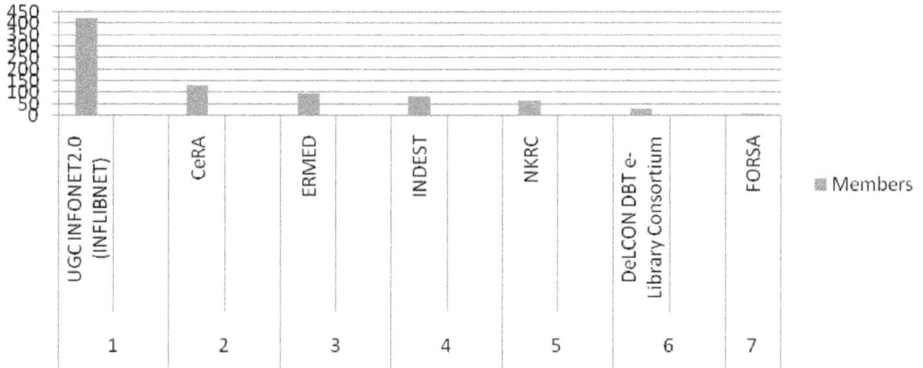


**Figure:1- Showing members of each consortia.**

With increases in prices of journals, shrinking library budgets and cuts in subscriptions to journals over the years, there has been a big challenge facing Indian library professionals to cope with the proliferation of electronic information resources. There have been sporadic efforts by different groups of libraries in forming consortia at different levels. The online information access system in India is had been past around two decades with the establishment of the Information and Library Network (INFLIBNET) Center. This one is a national Institute established by the University Grants Commission (UGC) of India in 1991. Now various subject-oriented consortia are established to improve the quality and status of research and educational development.

## 2.1. UGC-Infonet Digital Library Consortium (INFONET):-

The UGC-Infonet Digital Library Consortium was formally launched in December 2003. In this consourtium 419 universities libraries are utilising to provide online digital information services to their users. More than 7500 full text e resources are available from more than 20 international publishers to members universities and more than 5 types of bibliographical databases are availabe under the subcription. UGC-INFONET: e-journal consortium are also great boon to academic persons in the Indian universities. The journals subscribed from renouned publishers, compilers, aggregrators and society publications under the consortium, about more than 4000 full text scholarly electronic journals from around 25 publishers across the crountry can be accessed. Its provides currrent as well as archival access to core and peer-reviewed journals in different disciplines. This sconsourtium covered almost arts, humanities, social sciences, physical and chemical sciences, life sciences, computer sciences, mathematics and statistics other subjects areas are added time to time. This consourtium are sponsored by UGC and monitored by INFLIBNET(Information Library Network Centre) Ahmedabad.

## 2.2. CeRA e-journal consortium

The Consortium for e Resources in Agriculture is a targeted programme initiated by Indian Council of Agricultural Research in sub-project of NAIP, funded by World Bank. CeRA is an e-Consortium for Agricultural Libraries under National Agricultural Research System (NARS), established for facilitating accessibility of select scientific journals related to agriculture and allied fields, to all researchers in the (NARS). The subscriptions to journals by Libraries of ICAR Institutions and State Agricultural Universities (SAU) have been difficult due to financial crunch. To fulfill the requisition of the researchers and students the CeRA providing equal agricultural information access system facilities to all the agricultural university of states and CAU. The activities in CeRA are governed by Steering monitoring and negotiation and working committees and well supported by the Project Implementation Unit, NAIP. Its covered Deemed Agricultural Universities, ICAR Institutions, National Bureaus, Project Directorates, National Research Canter, SAU and CAUs. CeRA aims to develop the existing research and development information resource base of ICAR institutes / universities comparable to leading institutes/organizations in the world, to subscribe e-journals and create e-access culture among scientists/ teachers under the ICAR institutes/agricultural universities, and to study the impact of the consortium on the level of research publications measured through Science Citation Index (SCI) and NAAS Rating. CeRA members have access to on-lne journals through only IP authentication. It's providing completely full-proof system of security and also no need to memorizing user ID, password, and publishers User Resource Locator (URL).

## 2.3. Electronic Resources in Medicine in India (ERMED-India) Consortium

National Medical Library's Electronic Resources in Medicine Consortium is an initiative taken by Director General of Health Service & MOHFW to develop nation wide electronic information resources in the field of medicine for delivering effective health care. 39 centrally funded Government Institutions including 10 DGHS libraries + 28 ICMR Libraries and AIIMS library are selected at the initial stage as its core members. The MOHFW aims to provide fund required for the purchase of electronic journals under the NML-ERMED consortium project. The consortium will be coordinated through its headquarter set up at the NML. The nomenclature of the consortium is provisional; the same will be finalized after the approval of the competent authority. NML is already providing over 3380 print journal articles/ months across the country through postal service. Now efficiency of the same may be enhanced through electronic options are available to NML. 2068 print journals will be available for cross sharing. Package of 1350 online e-journals will be available from 12 leading publishers in the field of medical sciences. Free of charge 80 titles of Immunology & Microbiology + CIP will access complete 102 neuroscience collection + MEDLINE + Current Content database will be directly available to members. About 300 open access free journals will also be available.

The Consortium will purchase online e-journal from 12 different publishers. Users of e-journals have to visit website of different publishers to make use of full text. It may be time consuming and frustrating.

## 2.4. INDEST (Indian National Digital Library in Engineering Science & Technology)

The "Indian National Digital Library in Engineering Sciences and Technology (INDEST) Consortium" was set up in 2003 by the Ministry of Human Resource Development (MHRD) on the recommendation of an Expert Group appointed by the Ministry. The IIT Delhi has been designated as the Consortium Headquarters to coordinate its activities. The Consortium was re-named as INDEST-AICTE Consortium in December 2005 with the AICTE playing a pivotal role in enrolling its approved engineering colleges and institutions as members of the Consortium for selected e-resources at much lower rates of subscription. The Consortium enrols engineering and technological institutions as its members and subscribes to electronic resources for them at discounted rates of subscription and favourable terms and conditions. The Ministry provides funds required for subscription to electronic resources for 62 centrally funded Government institutions including IITs, IISc Bangalore, NITs, ISM, IIITs, IIMs, NITTTR's and few other institutions that are considered as core members of the Consortium. The benefit of consortia-based subscription to electronic resources is not confined to its core members but is also extended to all educational institutions under its open-ended proposition. 60 Govt./ Govt.-aided engineering colleges are provided access to selected electronic resources with financial support from the AICTE and 102 universities/institutions have joined the Consortium under its self-supported category in 2012. The total number of members in the Consortium has now grown to 1235. The INDEST-AICTE Consortium is the most ambitious initiative taken so far in India. It is the biggest Consortium in terms of number of member institutions in Asia. The Consortium attracts the best possible price and terms of agreement from the publishers on the basis of strength of its present and prospective member institutions. The Consortium subscribes to over 12,000 electronic journals from a number of publishers and aggregators. The consortium website at http://paniit.iitd.ac.in/indest hosts searchable databases of journals and member institutions to locate journals subscribed by the Consortium, their URLs and details of member institutions.

## 2.5. NKRC (CSIR-DST)

National Knowledge Resource Consortium (NKRC), established in year 2009, is a network of libraries and information centers of 39 CSIR and 24 DST institutes. NKRC's origin goes back to the year 2001, when the CSIR set up the Electronic Journals Consortium to provide access to 1200 odd journals of Elsevier Science to all its users. Over a period of time, the Consortium not only grew in terms of the number of resources but also in terms of the number of users as more like-minded institutes evinced interest to join the Consortium. Today, NKRC facilitates access to 5,000+ e-journals of all major publishers, patents, standards, citation and bibliographic databases. Apart from licensed resources, NKRC is also a single point entity that provides its users with access to a multitude of open access resources. The Consortium envisions emerging as a leader to serve the R&D sector with much needed information to strengthen the research and development system in the country.

## 2.6. DeLCON DBT e-Library Consortium

Department of Bio-technology e-Library Consortium (DeLCON) is a unique Electronic Journal Consortium, which is operational since January 2009. Currently the Consortium includes 16 DBT Institutions including ICGEB, New Delhi and 18 North Eastern Region (NER) Institutions. The Biotechnology Industry Research Assistance Council (BIRAC), New Delhi is also part of DeLCON. Now, the total 'DeLCON Members' are 34. A total of 926 selective Journals and a Database (SCOPUS) are covered under DeLCON. These all are accessible by the DeLCON Consortium Members through the DeLCON Portal (http://delcon.gov.in). Others can also view and access abstracts of papers as free of costs. DBT e-Library Consortium (DeLCON) is a unique Electronic Journal Consortium, which is operational since January 2009. Currently the Consortium includes 16 DBT Institutions including ICGEB, New Delhi and 18 North Eastern Region (NER) Institutions. The Biotechnology Industry Research Assistance Council (BIRAC), New Delhi is also part of DeLCON. Now, the total 'DeLCON Members' are 34. A total of 926 selective Journals and a Database (SCOPUS) are covered under DeLCON. These all are accessible by the DeLCON Consortium Members through the DeLCON Portal (http://delcon.gov.in). Others can also view and access abstracts of papers as free of costs.

## 2.7. FORSA (Forum for Resource Sharing in Astronomy and Astrophysics)

The oldest Library Consortium in India is the Forum for Resource Sharing in Astronomy and Astrophysics (FORSA) in subject of Physics was established in 1982. The types of consortia identified are generally based on various models evolved in India in a variety of forms depending upon the participants' affiliations and funding sources. Indian astronomy library professionals have formed a group called Forum for Resource Sharing in Astronomy and Astrophysics (FORSA), which falls under 'Open Consortia', wherein participants are affiliated to different government departments. This is a model where professionals willingly come forward and actively support consortia formation; thereby everyone benefits. As such, FORSA has realized four consortia, viz. Nature Online Consortium; Indian Astrophysics Consortium for physics/astronomy journals of Springer/Kluwer; Consortium for Scientific American Online Archive (EBSCO); and Open Consortium for Lecture Notes in Physics (Springer), which are discussed briefly.

# 3. Impacts of Library Consortia: -

3.1. **Financial benefits to the libraries-** the major factor involve in the establishment and update the library is financial support. The consortia are helping to the academic and research libraries to provide its user different types of e-books and electronic journals and other literature to full the user needs in a very less expenditure on electronic means.

**3.2. Established library cooperation:** - with the sharing of library resources as what ever the subscription from different publishers are available with the sponsored institute, its buildup a online connectivity with each other library and also a good relation among the library professional to share knowledge as well as their ideas and thoughts in providing a excellent virtual library and digital information utilization.

**3.3. Enhance Accessibility of library online resources by the users**: - the online accesses of information are very sophisticated and attractive and the library users are motivated through the consortia for regular utilization needful information available on the different portal of the library system. The consortia attract the library users to learn about e learning through library consortia and proper use of the soft copy of the information knowledge system.

**3.4. Quality library services: -** Library consortia are connecting the library user with the other electronic library services like Institutional repository, use of databases, CD Rom, E-books, OPAC and other kind of soft copy of literature. All such kinds of activities are indicating a best quality library service and showing complete automation of the library system and services.

**3.5. Enhancing reciprocal relation and leadership among library professional:-** Due to the library consortia a close relations are buildup among the library professional of different libraries and institutions have been seen and its create a leadership among the library to library and its creates a communication capacity among personnel and seen a great impact on the personality of library profession as it is a one kind of good job in human life.

**3.6. Impact on resource sharing:** - due to the library consortia the attendance of the users in the libraries are increasing and simultaneously the accessibility and use of the other printed documents are focusing an important among users to use the printed documents too. Specially when the needful information not available through consortia or on the www. Its provides communication capacity, logical discussion, expertise ness in access information and other many intangible knowledge.

**3.7. Impact on the quality of library service: -** The peoples and the authority of any academic institutions and research institution want to see the quality library service. The library consortia are a one kind of implication in the library service to share the knowledge information through information technology utilization. In this way the library consortia give a unique impact on the quality library service as traditional and modern library services as it is called hybrid library services that are completing users need with online access and with the printed version of document.

**3.8. Reduced the annual expenditure of the library: -** the consortia have reduced the annual expanses of the library in subscription of the periodicals as these are very costly and some time meagerness of library budget do not afforded to subscribed the foreign journals of reputed publishers like

Elgevier, John wieley, verlog, Taylor and Francis etc. in this connection the consortia helping the library specially in financial crunch.

**3.9. Increasing information accessibility: -** The consortia are a kind of motivation to the library users to access more and more tracing of information. The user is prone to access with the consortia and it's become a device to increase the users attendance and accessibility of information.

**3.10 Training and workshop opportunity to the staff and library user: -** The paradigm has been changed due to the application of ICT in the libraries and the consortium is a part of the paradigm shift in the library. Time to time training and workshop are organizes by the consortia sponsored institute for the other library client or sharer of the consortia. The library professional and user are benefited through it. And the library personnel become expertise to help to new comer students and researchers in the library.

**3.11 Enhancing the speed of research works: -** the researchers and scientists are able to get the related information in their research work and the references and reviews of literature are made available under the consortia sites. These are easier to tracing needful references for the researchers and students in sort of time and in less expenditure. And the speeds of research findings are concluded in a less time.

## 4. Conclusion

Besides of the above many other consortia are also running in the country as per need like ISRO, HELINET, ICMR, ICICI Knowledge Park, IIM Consortium etc. The online access through Internet system the consortia are playing a very crucial role in the higher education and research work. This creation has control the physical maintenance work of library professionals. The accessibilities of the research literature and their availabities as 24x7 have been increased. The time of the student and academic persons in searching of needful information save and directly took as soft copy in their devices to any time use. The different areas of subjects are created different portal of particular discipline information access. So there is no problem of interlinking of other subjects in each other. The benefit has been seen in the procurement of more electronic resources in the library in limited library budget. And this is what the libraries required in the present scenario. The libraries are preparing their own archives of soft copies collection of literature for future use. Now in this way the libraries are becoming very reach as their own repositories of knowledge, information. As required the particular piece of information can be taken printout.

There are some difficulties are seen against the print paradigm, the e-journal subscription and access models permit only licensing of the content or product for a period of stipulated time. Now the Soft copy cost is more in comparison to printed one. Now the online information access is an attractive information access and dissemination system for library and information services. The library consortia are full filling the main role of the libraries.

## References

Arora, Jagdish. (2001) Indian National Digital Library in Engineering Science and Technology (INDEST): A proposal for strategic co-operation for consortia-based access tdo electronic resources. International Information & Library Review, 33(2-3),149-165.

Babu, V. Nireekashana and Sivaprasad, G. (2010). Library consortia in India. Pearl: a Joural of Library and Information Science,4(4),220-223.

Chakravarti, R. and Singh, S. (2005). e-Resources for Indian universities: new initiatives. SRELS J.inf.Manage.,(42), 57-73.

Chandrashekharan, H.; Patle. Sarita; Pandey, P.S. et al (2012). CeRA- the e-Journal Consortium for National Agricultural Research System. Current Science,102(6), 847-851.

Evans, G.E. and Margaret Zarnosky Saponaro. (2012). Collection Management Basics: Sixth Edition. Library and Information Science Text Series, Libraries Unlimited. Oxford, England. (161-177).

Francis, A.T. (2005). Library consortia model for country wide access of electronic journals and databases. In TAV Murthy et.al (Ed.s) Proceedings of International Conference on Multilingual Computing and Information Management in Networked Digital Environment, Cochin.

http://delcon.gov.in

http://ejournal.niscair.res.in/index.phphttp://ejournal.niscair.res.in/index.php

http://paniit.iitd.ac.in/indest

http://ugcinfonet.jccc.in/about/about.asp

http://www.inflibnet.ac.in/publication/annualreport/AR-2009-2010.pdf

http://www.ncra.tifr.res.in/library/forsaweb/index.htm

http://www.nmlermed.in

Manish Kumar. (2010).Library Consortium: Panacea to Libraries Financial Crunch. In: Madan Kumar, Stanley, et al. (Eds). National Conference on Knowledge Management in the Globalized era: Lead papers, April 21-23, 2010, New Delhi, AALDI,247-253.

Mark, T. (2007). National and international library collaboration: Necessity, advantages. Liber Quarterly: The Journal of European Research Libraries, 17(3/4), 1–7.

Nabi Hasan. (2015).Best Practices in Library Consortia Management.In:Rawat,S.S.,etal (Eds). National on Integrating ICT in Agricultural Libraries in India: Policies, Issues and Challenges: Compendium of Agricultural Libraries & User Community, June17-19,2015, Izat nagar, AALDI ,222-227

Pathak, Sandeep. & Neela Deshpande. (2004). Importance of consortia in developing countries an Indian scenario. The International Information & Library Review, 36(3),227-231.

Rajoli, Iqbalahmad, U., Birdie, Christina, & Karisaddappa, C.R. (2005). Use of resources through consortia made in Indian Library & Information centers: A case study of FORSA Consortium. IASLIC Bulletin, 50(2), 74-82.

Sridhar, M.S. (2002) Resource sharing among ISRO libraries: A case study of consortia approach. SRELS Journal of Information management.39(1),41-58.

Thompson, T. L. (2004). Library consortia in the 21st century: Beyond the buying club. Trends in Law Library Management & Technology, 15(2), 1–4.

# 11

# Use of e-Resources and its Advantages

**Dev Walia**

Assistant Librarian, University Library
CSK Himachal Pradesh Krishi Vishvavidyalaya
Palampur -176 062, Himachal Pradesh
*email : devwaliacopi@gmail.com*

## ABSTRACT

*E-resources have changes the definition of libraries because it enhanced the collecting and organizing aspect of library. This has been became an important part of learning and teaching strategies in universities and colleges. Article contains benefits of e-resources for post-graduation students. The advent of Information technology gave possibility of production of information resources in different non-print formats. These are commonly called electronic resources of e-resources. They contain same information as their print counterparts but the information is in electronic format or digital format. They can be accessed and retrieval of information stored in them is possible with the use of computers or electronic devices.*

***Keywords:*** *e-resources, e-books, e-journals, e-database, Library, User community,*

## 1. Introduction

An electronic resource is any information that the library provides access to in an electronic format. The library has purchased subscriptions to many electronic information resources in order to provide user with access to them free of charge. The information resources could be print or electronic. E-resources, as defined by Ekwelem, Okafor & Ukwoma (2009), are information sources that are available and can be accessed electronically through such computer networked facilities as online library catalogues, the internet and the World Wide Web (WWW), CD-ROM databases etc. E-resources have the potential for enhancing postgraduate learning, as the resources provide postgraduates with vast quantities of information in an easily accessible non-sequential format. Postgraduates are major users of university library resources and services. This might be so because of their need for writing seminar papers, term papers, information for their assignments and other research activities. The presentation of a standard research work (thesis/dissertations) by postgraduates to their departments is the major component that will lead to the award of their final degrees. To present standard research work, postgraduates will need information resources. University libraries have the responsibility for providing a wide range of resources to meet postgraduate's research needs, collect information for their assignments and term papers, prepare for examinations and broaden their general knowledge.

## 2. Importance of e-Resources

E-resources become more indispensable after the arrival of Internet for a common man. Now the e-resources can be accessed over the internet without physically available in a library. They are hosted on publishers/aggregators/institutions servers and user can access them, search retrieved required information from them directly from respective website. Libraries have undergone drastic change in recent years, in both information gathering and dissemination techniques. Electronic collections are helpful especially to distant learners who have limited time to spend in the library. These days users can access the library via off campus access. Libraries are no longer operating in a purely physical environment, but more often in a mixed environment that consists of both print and e-resources. With the development of the internet and the wealth of e-resources now available on shelves in the library, but electronically on CD-ROM (Compact Disc, Read Only Memory), DVD (Digital Video Disc) or the internet. Advancement in technology have changed the way information is presented and circulated. Therefore the collection of the libraries not limited to print collections that require users to visit the library.

## 3. Types of E-Resources

Different types of e-resources mostly used by the postgraduates in their libraries are:-

- e-books
- e-journals
- e-newspaper

- e-magazine
- e-research reports
- e-dictionaries,
- e- encyclopedia
- e-thesis
- Online databases etc.

Databases are used to refer to indexing and abstracting databases or called aggregated databases which are essentially collections of full-text electronic journals in libraries.

Electronic journals are journals that exist in digital form on the internet, or on CD-ROM or DVD. They can be either an electronic version of a print journal, or exist only in electronic format.

Electronic resources have been used to encompass all information sources (books, journals, databases and other formats) that exist in an electronic/digital format, e.g. computer discs, CD-ROM, DVD or accessed on the internet. E-resources are those resources which include documents in electronic or e-format that can be accessed via internet in digital library environment. E-resources are that electronic product that delivers a collection of data, be it text, image collection, other multimedia product like numerical, graphical mode for commercially available for library and information centre's.

## 4. Advantages of E- Resources

- International Reach
- Any type of data or material regarding any thesis or synopsis is available in e-resources for PG students.
- Availability for users.
- Space saving properties
- Immediate access of information
- Great speed of communication
- Unlimited capabilities
- Convenience
- Enhanced search ability
- Linking

## 5. Disadvantage of E- Resources

- It needs internet connection. Slowness of connection is also a drawback.
- Subscription to material in electronic form is more expensive than subscription to material in traditional form.
- A major drawback of electronic resource is comprehensiveness. Electronic resources do not generally date back as far as their printed counterparts.

## 6. Preference of e-Resources than print Resources

Electronic resources along with printed sources have become an integral part of library collection. A large number of electronic resources have acquired by libraries. Students use e-resources to prepare seminar/project, to prepare for their examinations, to update information. E-resources are easier and faster access to information. These can access to a wide range of information. Various other purposes for a user to use e-resources includes: gathering information on a specific topic, gaining general information, obtaining answers to specific questions, completing assignments, reviewing literature, writing essays and helping decision making. The use of e-resources made end user, saved time and money. A large majority of students use online journals for preparing project reports and for listing references only.

Other advantages of electronic information resources include multiple choices of formats, provides faster & reliable communication, multi user access, user sitting on their desktop can access e-resources.

## 7. Problems faced by the users

The most common problem with e-resources are low speed connectivity, lack of awareness about statutory provision for accessing e-resources by the institutions, technical problems (software/hardware), unavailability of sufficient e-resources, doubts in permanency, high purchase price, lack of legal provision. Other problems are download delay, failure to find information, lack of search skills, high cost of access, power outages, unavailability of some websites and difficulties in navigating through e-resources.

## 8. Conclusion

If we put advantages and disadvantages of electronic journals on one scale, certainly there are more advantages. We can say that e-resources have changed the definition of libraries because it enhanced the collecting and organizing aspect of library. It has changed the way of seeking and disseminating information. Students, researchers use online e-resources for their need. This has been became an important part of learning and teaching strategies in universities and colleges.

## References

Selvaraj, A. D., and G. Rathinasabapathy. "A study on electronic information use pattern of faculty members of self-financing engineering colleges in Tiruvallur District, Tamil Nadu." Asian Journal of Library and Information 6 (2015): 3–4.

Selvaraj, A. D., and G. Rathinasabapathy. "Information Use Pattern by Students of Self-financing Engineering Colleges in Tiruvallur District, Tamil Nadu: A Study"." Journal of Library, Information and Communication Technology 6.3-4 (2015): 45–54.

Shehbaz Husain Naqvi, (2014). The use of e-resources by postgraduate Engineering Students at Jamia Millia Islamia University. Paripex-Indian Journaj of Research, volume-3, issue 7.

Walia, Dev (2014). Information seeking pattern and information seeking behavior of postgraduate students of CSKHP Krishi Vishvavidyalaya in digital era. In : Pandey, Mathur et al (Eds) Agricultural Information management in Digital Era : national

Conference of Agricultural Librarians and user Community : Lead papers February 4–5, 2014, Raipur.

Walia, Dev. (2012). A case study of CSKHPKV University Library. In : Jain, Arun kumar et al. (Eds) Role of Agricultural Libraries in knowledge Management : National Conference of Agricultural Libraries and User Community : Lead papers March 15-16, 2012, Hyderabad. 407–413.

Walia, Dev. (2015). Use of CeRA (Consortium for e-Resources in Agriculture) by the post-Graduate students of CSKHP, Krishi Vishvavidyalaya, Palampur : A Case Study. In: Rawat, S. S. et al (Eds) Integrating ICT in Agricultural Libraries in India : Policies, Issues and Challenges : Lead papers June 17–19, 2015, Izatnagar.

# 12

# Goods and Services Tax Introduction Process: An Overview

**Ambhore Sagar Pandit[1] and Dr. Khaparde Vaishali[2]**

[1] Research Scholar; [2] Professor and Head Department of Library and Information Science
Dr. Babasaheb Ambedkar Marathwada, University Aurangabad, Maharashtra State
e-mail: khapardevaishali@gmail.com

**ABSTRACT**

*The Goods and Service Tax (GST) is one of the biggest taxation reforms in India, the decision on which is pending. The central idea behind this form of taxation is to replace existing levies like value-added tax, excise duty, service tax, and sales tax by levying a comprehensive tax on the manufacture, sale and consumption of goods and services in the country. GST is expected to unite the country economically as it will remove various forms of taxes that are currently levied at different points. The Present study discusses about, Need of GST, Concepts and Meaning of Goods and Service Tax, Who would be impacted, Applicability to service providers, Time to Plan for GST, Taxable event, GST collection model, Continuation of exemption currently available, Salient features of the GST model.*

***Keyword:*** *- Goods and Service Tax, GST, GST collection model, Time to Plan for GST.*

## 1. Introduction

Tax policies of a country play an important role on the economy through their impact on both efficiency and equity. A good tax system should keep in view issues of income distribution and at the same time, also generate tax revenues to support government expenditure on public services and infrastructure development. The framework of value added tax (VAT), recognized as GST as well in several countries, has been one of the major development in taxation structures worldwide. More than 135 countries adopted the GST/ VAT framework effectively. Indian economy is getting more and more globalised. Introduction of an integrated Goods and Services Tax (GST) to replace the existing multiple tax structures of Centre and State taxes is not only desirable but imperative in the emerging economic environment. The implementation of GST would ensure that India provides a tax regime that is almost similar to the rest of the world. It will also improve the international cost competitiveness of native goods and services. According to a report by the National Council of Applied Economic Research, GST is expected to increase economic growth by 0.9 percent to 1.7 percent and Exports are expected to increase by 3.2 percent to 6.3 percent.

## 2. Need of GST

In case of service tax we have the CENVAT & by case of VAT we have input tax credit to reduce the cascading effects which was present in case of central sale tax system. But there are goods & services which are exempted at state & central level. Where the tax is not taken, no credit is provided for taxes which are paid on inputs & this causes cascading of taxes. In case of VAT Agriculture, real property sector, gas & oil production, mining & entire service sector are exempted from VAT, therefore no credit would be allowed to theses sector. One of the major reasons of cascading of taxes is central sales tax due to above reason there is a break in chain of set off which causes cascading of taxes. In case of service tax, there is a long list of services on which tax to be charged when services are provided in combination with goods then it is very complex to calculate service tax. E.g. works contract in this case we provide goods as well as services. In services tax we charge tax on services only, so the decision of percentage of goods sold & services provided we depend upon the complex method which is given in the service tax rules therefore for these kinds of services, payer needs to know the detailed provision of service tax.

## 3. Concepts and Meaning of Goods and Service Tax

The Goods and Services Tax (GST) is a comprehensive value added tax (VAT) on the supply of goods or services. France was the first country to introduce this value added tax system in 1954 devised by a public servant. In India, due to non consensus between central and state government, the proposal is to introduce a Dual GST regime i.e. Central and State GST.

## 4. Who would be impacted?

All businesses, whether engaged in sales / supply of goods or supply of services, would be impacted by GST. The impact would be on supply chains, ERP, product pricing, dealer margins etc.

## 5. Applicability to service providers

Unlike the transition from the sales tax regime to the VAT, where only businesses dealing in goods were affected, in the case of GST, as the name suggests, both goods and service providers will be impacted. Thus, even pure service providers need to plan for the transition to the GST.

## 6. Time to Plan for GST

The draft laws will clarify finer aspects of GST such as rates, classification and compliances. However, based on the material in the public domain, one can begin with spreading awareness among various stakeholders within the organization and identifying broad areas of action before the draft laws are published. Experience of VAT implementation suggests that there may not be enough lead-time available between the date of announcement of GST implementation and the actual date of GST implementation.

## 7. Taxable event

The Taxable event will be the supply of goods and the supply of services. Hence, the current taxable events such as 'manufacture of goods', sale of goods' and ' rendition of services' will not be relevant under the GST regime.

## 8. GST collection model

GST is collected on the value added at each stage of sale or purchase in the supply chain. The tax on value addition is ensured through a tax credit mechanism throughout the supply chain. GST paid on the procurement of goods and services is available for set-off against the GST payable on the supply of goods or services. The idea is that the final consumer will bear the GST charged to him by the last person in the supply chain. It is thus a consumption based indirect tax.

## 9. Continuation of exemption currently available

In view of the fact that under the GST scheme, exemptions would be minimal, it may not be correct to proceed on the assumption that the present exemptions would continue under the GST dispensation.

## 10. Salient features of the GST model

Keeping in view the report of the joint working group on goods and services tax, the views received from the States and Government of India, a dual GST structure with defined functions and responsibilities of the Centre and the States is recommended. An appropriate mechanism that will be binding on both the centre and the states would be worked out whereby the harmonious rate structure along with the need for further modification could be upheld, if necessary with a collectively agreed constitutional amendment.

## 11. Conclusion

GST is an improvement over VAT & service tax but it is also having some flaws. Introduction of the Goods and Services Tax (GST) will be a significant step towards a comprehensive indirect tax reform in the country. It is expected to bring about efficiency and transparency in the indirect tax mechanism in India. Further it will also encourage an unbiased tax structure that is neutral to business processes and geographical locations.

## References

Aaron Henry J., ed (1981). Value Added Tax: Lesson from Europe. Washington DC: Brooking Institution.

Adams, D.W. (1980). "The Distributional Effects of VAT in the United Kingdom Ireland, Belgium and Germany." The three Review 128 (December).

Anjanaiah M, (1971). "Do excise duties aggravate inflation? Economic and Political Weekly (Feb.18)

Atikson, A. B. and J.E. Stiglitz. (1990) Lectures on public Economics. New York: McGraw Hill.

Aujean, Michel, Peter Jenkins and Satya Poddar (1999): "A New Approach to Public Sector Bodies", 10 International VAT Monitor 144 (1999).

Bagchi, Amaresh et al (1994), "Reform of Domestic Trade Taxes in India: Issues and Options", National Institute of Public Finance and Policy, New Delhi

Narayana, A V L, Amaresh Bagchi and R C Gupta, (1991), " The operation of MODVAT" – Vikas publishing House Pvt. Ltd.,

# 13

# Information Literacy Programme

**Ranveer Vishakha B[1] and Dr. Khaparde Vaishali[2]**

[1] Research Scholar [2] Professor and Head
Department of Library and Information Science,
Dr. Babasaheb Ambedkar Marathwada University
Aurangabad, Maharashtra State
*e-mail: khapardevaishali@gmail.com*

**ABSTRACT**

*Information literacy (IL) is the set of skills that allows us to find, evaluate and use the information we need. Libraries have long been involved in training their users in library use, its services and sources. Programmes such as library orientation, user education, etc organize by libraries to teach the user how to use library, how to search information sources and how to access information in the digital age. Information literacy is critically important in present age, because we are surrounded by a growing ocean of information in all forms. Research in this area in developing countries such as India is still in primary stage. The Present study discusses about, Use of Information and Communication Technology, Significance of Information Literacy, What is Information, The Need For Effective Use of Information, Why is Information Literacy So Important and Who Needs Information Literacy.*

***Keywords:*** *Information literacy, information literacy programmes.*

## 1. Introduction

Information Literacy is the vital process in the modern changing world which is mostly used for higher education particularly, at the university level education. In our information-centric world, students must develop skills early on so they are prepared for post-secondary opportunities, whether in the workplace or in pursuit of higher education.

Information literacy "enables people to interpret and make informed judgments as users of information sources, as well as to become producers of information in their own right. Information literacy means information awakening in person about the needed information. Information literate people are able to access information about their health, their environment, their education and work, empowering them to make critical decisions about their lives, e.g. in taking more responsibility for their own health and education".

## 2. Definition of Information Literacy

The American Library Association's (ALA) Presidential Committee on Information Literacy, Final Report states, "To be information literate, a person must be able to recognize when information is needed and have the ability to locate, evaluate, and use effectively the needed information" (1989).

## 3. Use of Information and Communication Technology

The modern world is known as Information Technology. Information technology skills enable an individual to use computers, software applications, databases, and apply related technologies to achieve a wide variety of academic, work-related, and personal goals. Among these are information literacy focus on content, communication, analysis, information searching, and evaluation; whereas information technology "fluency" focuses on a deep understanding of technology and graduated, increasingly skilled use of it. For the effective teaching and learning we use demos, tutorials, course materials in electronic form as interactive and ordinary CDs, animated and multimedia programmes running on cable TV network, FM, All India Radio, local Newspapers that attract the learners.

## 4. Significance of Information Literacy

Information Literacy is a transformational process where the learner needs to find, understand, evaluate, and use information in various forms to create for personal, social or global purposes. Information literacy is a set of abilities requiring individuals to recognize the information with the ability to locate, evaluate, and use effectively. (Gilton, 1994) expressed Information literacy cannot be equated with computer literacy as it requires a technological know-how to manipulation of both computer hardware and software or library literacy which in turn requires the ability to use a library's collection and its services, although there is a strong relationship among all these concepts. Each of the literacy requires some level of critical thinking. But compared with computer literacy, information literacy goes beyond merely having access to and knowledge of how to use the technology because technology alone does not guarantee quality learning experiences. Compared with library literacy, information literacy is more than searching through an online catalogor other reference materials because information literacy is not a technique rather it is a goal oriented skill for learners. However, (Shapiro & Shelly,1996) have defined Information literacy as a new liberal art that extends from knowing how to use computers and access information to critical reflection on the nature of information itself its technical infrastructure and its social cultural and philosophical context and impact.

## 5. What is Information?

Information is a resource that has varied definitions according to the format, and media used to package or transfer it, as well as the discipline that defines it. (Case,2002) provides a broader definition. Here the term is synonymous with:

- Encapsulated knowledge
- Packaged human experience
- A source that can provide a myriad of data
- A resource that takes different formats, packaging, transfer media, and varied methods of delivery
- People: family, friends, tutors, fellow students
- Institutions, i.e., national health service professionals or help facilities

## 6. The need for effective use of Information

Information has become a vital source for world economies and is certainly the basic component of education. Information is a vital element to technological and scientific change. It poses several challenges to individuals of all walks of life: students, workers, and citizens of all types. The current information overload requires people to validate and assess information to verify its reliability. Information by itself does not make people information literate. Information is certainly a:

- A vital element for creativity and innovation
- A basic resource for learning and human thought
- A key resource in creating more knowledgeable citizens
- A factor that enables citizens to achieve better results in their academic lives, with regard to health, and at work
- An important resource for national socio-economic development.

## 7. Why is Information Literacy so important

Information literacy is important owing to the amount of information that is available in contemporary society. Simply being exposed to a great deal of information will not make people informed citizens; they need to learn how to use this information effectively, (ACRL, 2000). Data Smog refers to the idea that too much information can create a barrier in our lives. Especially students and the society require a special skill to handle this fast increasing information, in order to use their educational and economical purposes more effectively. Information literacy is considered as the solution for the data smog. (ACRL ,2006). Information literacy allows us to cope with the data smog, by equipping us with the necessary skills to recognize when we need information, where to locate it, and how to use it effectively and efficiently. Consequently it will help decision making and productivity which is beneficial to the society.

## 8. Who needs Information Literacy

Information literacy skills are helpful to everybody, especially students, in order to succeed academically and in their future job opportunities. Teachers and lecturers are greatly in need of information literacy skills, in order to carry out their occupations efficiently and successfully. Basically, everybody in the society is in need of information literacy skills. Information literate individuals improve the society's quality of life in general and academically. Information literacy helps us in our day to day life such as buying a house, choosing a school, making an investment, voting for the election, and many more. Information literacy skills are of prime importance in order to achieve every body's academic goals. Truly information literacy is the foundation of the democratic society.

## 9. Conclusion

Information literacy which is a transformational process lies at the core of lifelong learning and it empowers people in all walks of life to obtain, evaluate, use and generate information effectively so to achieve their personal, social, occupational and educational goals. Further in a democratic setup, it is also a fundamental human right in a digital environment that promotes social inclusion of all nations. In the process, information literacy enables sharing of resources, knowledge more effectively with the help of technology.

## References

ACRL (Association of college and research libraries) (2000) Information Literacy, Journal of Documentation, No. 57, pp. 218–259.

Bawden, D. (2001, March). Information and Digital Literacies: A Review of Concepts. In Nov 2014] http://www.ala.org/ala/acrl/

Breivic P S& Senn, J. A. (1998) Information literacy: educating children for the 21rst century. 2nd Ed.).Washington C, National Education Association,

Bruce, C S. (2002), Information Literacy as a Catalyst for Social Change: A Background Paper. White Paper prepared for Unesco, the U.S. National Commission on Libraries and Information Science, and the National Forum for Information Literacy, for use at the Information Literacy Meeting of Experts, Prague, The Czech Republic. Available at: http://www.nclis.gov/libinter/ infolibconf&meet/papers/ bruce-fullpaper.pdf

Bruce, C. (1997). The Seven Faces of Information Literacy. Seven Faces of Information Literacy. AULSIB Press, Adelaide Auslib Press.

Doyle, C S.1992. Outcomes Measures for Information Literacy within the National Education Goals of 1990 .Final report to National Forum on Information Literacy, summary and findings.

Grassim, E. (2004) Information Literacy: Building on Bibliographic Instruction. American Libraries, vol 35, no 9, pp51–53

# 14

# Social Media and Libraries: Emerging Opportunities and Challenges

**Dr.S.Kopperundevi**

Assistant Librarian, Veterinary College and Research Institute
Tamil Nadu Veterinary and Animal Sciences University
Tirunelveli – 627 358, Tamil Nadu
*e-mail : devielangowins@gmail.com*

## ABSTRACT

*Person to person communication destinations turn out to be more prominent with clients of today's groups. In this information revolutionized era, libraries are subjected to improve the administrations and items with the propelled Information Technologies and Applications, for example, Twitter, Facebook, Instagram, and YouTube permit the users to eagerly take an interest in what is being written and published. These sites have made a domain where Librarians can achieve their users in their classrooms, homes, on their jobs, or even on furlough. So many libraries are incorporating their services into existing person to person communication situations. It is a way for the library to stay connected with the users outside of a confined building. The main objective of this paper is to describe the use of social networks in the library services. SNS also called virtual community or profile site. This paper describes the types of social networks available and how libraries can adopt it in a meaningful and purposeful way. The challenges faced by LIS professionals in this context are also briefed.*

***Keywords:*** *Social Media, Social Networks, Libraries, Twitter, Facebook, Instagram, YouTube, Social Networking Site*

## 1. Introduction

Boyd and Ellison defines "Social networking sites as web-based services that allow individuals to construct a public or semi-public profile within a bounded system, articulate a list of other users with whom they share a connection, to view and navigate their list of connections and those made by others within the system".

Social networking sites (SNS) are new developing technology in sharing and disseminating information product with in the users. SNS provide an innovative and effective way of connecting users. Social networking web sites promote a number of social network services. The use of online social networks by libraries and information organizations is also increasingly prevalent and a growing tool that is being used to communicate with more potential library users, as well as extending the services provided to individual libraries. It allows users to share ideas, events, activities, and interests within the individual networks. SNSs, such as Friendster, LinkedIn, MySpace and Facebook, set up personal communities, allow users to make persistent comments on the profiles of their friends and send private messages. These features make SNSs excellent in initiating interaction among users. The number of libraries which adopt SNSs is increasing. In a 2009 survey, researchers found that SNS was only adopted by a few academic libraries. After 2 years another survey revealed that Facebook and Twitter have become the most popular Web 2.0 applications in libraries. Now most of the libraries are using. SNS serves as the best tool to identify and serve to the information seeking community. Where we can get the feedback on library products and services, with the help of support forums and can discuss and resolve the issues. It mainly allows users to share tips, resources and also to promote the quality products and services of Library and information centers.

## 2. Need of Social Networking Tools in Libraries

Libraries can use social networks for a variety of purposes

- To permit users to recommend the purchase of documents or give a comment to the library
- To show Image galleries of library events
- To make use of the information resources
- To teaching users, for the reference service
- To encourage services
- To spread the information
- To communicate with the library users

## 3. Types of Social Networking Sites (SNS)

The below mentioned are social networking tools which can be used for the purpose of promoting the library services. SNS allow librarians to adopt a new role by placing themselves into a social realm with users.

**aNobii:** Social networking site like aNobii helps book lovers to share reviews and recommendations. It also prepare due date alerts, lending, and discussions.

**Bebo:** A popular social networking site where users can share photos, stories, their journal, and more with loved ones secretly or openly on the Internet.

**Blog:** Blogs have several potential uses by academic libraries. By creating a blog, librarian will be able to disseminate information to lots of people at one time. Librarian can update students on new collections, or just conversing with library staff, blogs are a powerful tool, especially when combined with RSS. Blogs encourage user interaction through their comment feature, which allows students to provide feedback regarding the information provided and the library itself.

**Classmates:** One of the largest and most used websites that brings together and allows people who graduated from high school and allows you to keep in touch with them and any future reunions.

**Community Walk:** Community Walk offers a geographical way to interpret text and events. Librarian can use it for instruction, such as showing someone where to find a book, or walk them through a historical and geographical timeline.

**Connotea:** This is a free online reference management service tool where librarian can save, organize links to their references and can share references with their users. It will be useful for researchers and scientists.

**Daft Doggy:** If you've found a particularly good resource, you can use Daft Doggy to create a website tour with instructions, pointing out useful references and items of note.

**Del.icio.us:** It offers a way to store and organize bookmarks (by tagging) and allows to share with others. It is known as a social bookmarking tool also.

**Digg:** Digg is a great way to find useful content that you wouldn't come across in traditional ways. Find stories here, then share them with others using Digg's blog function.

**Facebook:** Present day Facebook is a popular destination for users to setup their own personal web pages, connect with friends, share pictures, share movies, talk about what you're doing. It is most popular now because it is librarian-friendly, with many applications like World Cat, and much more. Librarians can interact with users to know their information need. Libraries try to link some of these specialized library applications to Facebook. They are publishing day today news, exhibitions, providing reference services and upcoming sharing workshop information, conducting online quiz's and week end programs, creating public awareness program about libraries.

**Email-Facebook:** offers a fully functional internal email system that also manages messages, friend requests and other connections.

**Facebook chat:** Facebook enables to chat with other facebook user from the facebook page.

**Facebook groups:** FaceBook enables to connect to people with liked interests.

**Flickr:** Flicker is the largest Image sharing website in the world. Librarians can use this tool to share and distribute new photos of library collections, conferences, program organized within campus. Cover page of new arrivals of both books and journals can be disseminated to users.

**Friendster :** A popular social network that brings together friends, family, and allows you to meet new people who share similar interests to you from all over the world.

**Podcasts**: Audio files available for download via subscription, so it can be automatically downloaded to a computer or MP3 player.

**Footnote:** On Footnote, you'll get access to original historical documents, and can update them with your own content and insights. You can even find personal anecdotes and experiences you won't find in reference books.

**Frappr:** Librarian can share their library map using this social website.

**Gather.com:** This website is use to share articles about libraries, pictures and also to discuss various issues through group discussion.

**Goodreads:** This is a social cataloguing website. Users can freely search Goodreads extensive user populated database of books, annotations and review.

**Google +:** Used to Connect with other Google+ users. Librarian can create the own profile & make connection with other library professionals to create the own profile & make connection with other library professionals

**Google groups:** Google groups at the time of writing the author is not aware of reference librarians using Google groups although such a staff group was formed for general discussions and exchange of information regarding Library 2.0.

**Instagram:** Instagram is an online photo-sharing, video-sharing and social networking service that enables its users to take pictures and videos, apply digital filters to them, and share them on a variety of social networking services, such as Face book, Twitter, Tumblr and Flickr. Libraries are doing lots of cool things on Instagram. Libraries posted pictures of staff, patrons, shelves of books. Libraries are sharing bookface pictures where people pose with covers of books that line up with their faces. They highlight their programs and facilities. They advocate for literacy and community.

**Italki.com:** Italki is a language learning social networking tool. Librarian can be linked with library website for students to learn various languages.

**Library Thing:** Social cataloguing web application for storing & sharing book catalogue & various types of book metadata, can catalog along with Amazon, the Library of Congress, and more than 200 other libraries around the world. You'll get recommendations and easy tagging as well. Librarians can utilize this to send a list of current publications to users.

**lib.rario.us:** Another social cataloging site, you can put media such as books, CDs, and journals on display for easy access and tracking

**LinkedIn:** LinkedIn is an interrelated professional network of skilled professionals from about the world, LinkedIn is for the most part for curtailed the professionals, users to preserve a list of contact information of persons. This social networking site for professionals is a great way to get library patrons connected with the people that can help them find information. Whether that's you, faculty, authors, historians, or other sources, they can find them in your LinkedIn network

**Lislins:** Professional site for LIS students & Librarian , used to share & implement new ideas in the library Profession.

**Meebo:** Network and assist students on Meebo, no matter what IM client they use. Online chatting or virtual reference service in library can impacted by professionals to users.

**Microblog:** A blog that is made up of short posts of 140 characters or less.

**MySpace:** MySpace is one of the most popular and fastest growing social networking sites with millions of users using this site and one of the most viewed website on the Internet. Every registered user can share individual profile, photos, videos, etc. librarian can take advantage of this site to post, calendar, custom catalog search tools, and blog features to improve their presence, they can share their ideas and thoughts to the users

- Search and browsing- Myspace enables to search and browse other users by name, email, address, age, gender, location, education, or level of interest.
- Email- Myspace offers a fully functional internal email system that manages messages, friend request, and other connections.
- Instant Messaging-Myspace enables to download an instant messaging program that lists all Myspace friends without requiring that log on to Myspace.

**Ning:** Librarians can get connected with users, library associations, and more. Librarian can utilizes this platform to discuss topical issues among the members.

**Netvibes:** In Netvibes' new Ginger beta, you can create a public page that can be viewed by anyone. You can use it to help guide patrons to helpful internet sources, news feeds, and more. It can be integrated with many of the tools mentioned here, like Flickr and library blogs.

**Orkut:** A popular service from Google that provides you a location to socialize with your friends and family, and meet new acquaintances from all around the world.

**Pinterest:** An upcoming and popular picture and sharing service. Pinterest allow users to save images and categorize them on different boards. They can follow other users' boards if they are likely minded. Pinterest users can upload, save, sort and manage images, known as pins, and other media content through collections known as pin boards.

**RSS feeds:** RSS stands for Really Simple Syndication and Rich Site Summary. RSS is a way for subscribers to automatically receive information from blogs, online newspapers and podcasts, etc. It allows you to easily stay informed by retrieving the latest content from the sites you are interested in. RSS can be applied to some of the following Library & Information Services: Selective Dissemination of Information, Current Awareness Service, Bibliographic Service, Bulletin Board Service.

**Second Life:** On Second Life, librarian can create a virtual library with streamed media, discussions, classes, and more.

**Slide share:** It helps to share the power point presentation. Librarian Get upload the best PPT's of the students & can be linked to library website and college website for future reference. It's a great way to disseminate information among research community to the field of research and development (R&D) activities.

**Students Circle Network:** This website connect students, teachers & institutions to course resources, study groups and learning spaces. Useful for teachers and students to study online, discuss the issues related with study & share information, librarian can link this website in the library website.

**Stumble Upon:** Another way to find great content is with Stumble Upon. Users can channel surf the Internet to find useful content, research tools, and more. Users who vote for web pages they like and dislike and allow users to create their own personal page of interesting sites they come across.

**Tagging:** Refers to the ability to add subject headings to content in order to organize information in a meaningful way and to connect to others that tag similar content in the same way.

**TeacherTube:** TeacherTube, which is a YouTube for teachers, presents an excellent opportunity for instructor-librarian collaboration. Instructors can guide students to helpful library resources, and vice versa.

**Teachstreet:** It is deals with Education/ Learning/ Teaching - More than 400 subjects. It can be linked with college websites where students can get information about subjects.

**Tumblr:** Librarian can post text, links, quotes, audio clips, videos and images to a community of followers and outside reader.

**Twitter:** Twitter is a social network and micro-blogging service it permitted users to post their latest updates. Librarians can create twitter account for their own library where users can followers and librarian can tweet the latest news, updates events etc. on tweeter account. Twitter use for academic conversation, had a positive effect on students grades, engagement and inspiration.

Twitter is the fastest rising social networking site in the world. Profiles can be customized, both colors and background image

- Users can update their status.
- Users can repost another users status update, using the ReTweet function.
- Powerful searching based on users replying to each other and tagging of tweets.

**YouTube:** This social networking that permitted users to post their own videos. Anyone can watch and sharing videos, whereas registered users are permitted to upload an unlimited number of videos. A range of case studies of library use of many social network tools can be set up YouTube. Library video and e-learning tutorials, events and others video library services can be effectively promoted and webcast through YouTube. And also libraries have been using YouTube as library virtual tour guides. Users can download or view library tour guides via YouTube and learn themselves on how to use library better for the studies and research activities.

**Virtual worlds:** Allows for real-time communication and collaboration from all over the world. Each person uses an avatar as a virtual representation of themselves.

**Wikipedia:** A collaborative space for developing web content. No web design knowledge is needed to create a wiki. It is an online encyclopedia updated by users. Librarian can use this tool to share your knowledge by editing, or simply point library patrons in the right direction. Wikipedia: Wikipedia is an online encyclopedia updated by users. Librarian can use this tool to share your knowledge by editing, or simply point library patrons in the right direction. Academic libraries can create subject Wikis with links to resources on a chosen topic or for a particular class, including information regarding relevant databases and search tips tailored to that subject. Students conducting research on a topic can use the resources provided as well as edit the Wiki to include additional information. Thus, a Wiki-based subject guide allows for collaboration between academic librarians and the students.

**Yik Yak** – Smartphone social network that connects users who are in close to each other.

## 4. Challenges faced by LIS Professionals

Lack of adapting the new technologies creates a great challenge. Challenges associated with using social media in libraries include the following:

**Lack of knowledge:** Librarians in the developing countries are not aware of social networking services

**Bandwidth problem:** Most institutions have restricted bandwidth to bolster this practice

**Technophobia:** Many librarians and users are scared of handling computers and networks

**Lack of digital preservation culture:** Preservation of digital collection is seriously lacking in most institutions in developing countries.

**Untrustworthy power supply:** The low supply of electricity discourage people from taking an interest in the online forum.

**Lack of training of staff:** Most librarians do not have the new IT skills that could be required to adopt the social networking tools for effective library services.

**Government intervention:** There is little or no intervention of the government in the area of ICT.

**Technological expertise:** Social media can require technological expertise, for example customizing applications to provide access to online catalogs.

**Social media content:** It can be a challenge for librarians to use an informal but presentable tone, or deliver social media content in a bilingual or multilingual region.

**Budget constraints:** There are limited funds to support more advanced social media usage/features and the training that would be required to enable this.

**Librarian should Work hard:** Library needs to work hard to maintain engagement with library users and attract popularity.

**External factors:** External factors such as Internet connectivity, technological infrastructure and government restrictions on the use of social media may restrict access.

**Culture:** The need for an open, transparent, horizontal working culture. It is not always a prerequisite but it is conducive for effective and creative online knowledge sharing.

**Organization Support:** To have a commitment from the management for collaborative web tools.

**Conviction:** Having good arguments to proof why these tools are useful.

**User Orientation:** Developing a web-based communication culture needs orientation.

**User Participation:** In the beginning usually only few users participate; that's why a critical mass of contributors is important.

**Resources:** Be aware the tools are cheap and easy to install, but do not underestimate the resources you need

**Need Patience:** To incorporate web2.0 tools to an organization takes time.

**Training:** Many people from the organization are totally new to the applications.

**Usability:** Usability is very important because users shall take advantage of all features offered. For example many wikis especially lack usability.

**Software:** Implement a solution on your own server or rely on an application service provider.

## Privacy issues, IPR, copyright issues of social software

**Technical and institutional barriers:** Certain social software may conflict.

**Organizational policy:** Libraries need to focus more on managerial arrangements while utilizing the social networking sites. Libraries must be sure about dealing with the clients reactions and giving the rules. Libraries can utilize the applications to advance the occasions or library benefits yet can't be altered as they need.

**Absence of obscurity:** SNS for the most part asking name , area ,sexual orientation and numerous different sorts of individual data Time .

**Consuming:** Actual searching of data may be diverted.

**Absence of clarity:** Many users uses phrases & Short form which can be confounded.

**Absence of nonverbal communication:** They may also be difficulties in interpersonal communication when the messages are not clear.

## 5. Conclusion

Social networking site gives incredible chances to Library experts to interface with their clients as it spots them in the advanced social space of their clients. Library experts can get direct data about the client through collaborating with them. Library professionals can get first hand information about the user through interacting with them.

Social networking sites would help to set up a genial relationship in the middle of Librarians and Users; showcase the library's administrations. Person to person communication with clients, additionally help the free sharing of data that is searchable to general society or individuals from an informal organization. Present patterns bring up that long range interpersonal communication locales are essential to the lives of users. Lastly, users need to be quantitative and qualitative research about the use of Social networking sites tools as a form of student outreach to determine its effectiveness within libraries. Social networking sites can also be used as dynamic service and librarians take an active responsibility in developing.

## References

Anand Y. K (2015). "Importance of Social Networking Sites for Libraries and Information Centers". International Journal of Innovative Knowledge Concepts, 1 (2) p.29

Arumugam J and Rathinasabapathy, G (2013). "Social networking sites: a boon for libraries". Asian Journal of Library and Information Science 5 (3-4) p.24

Boyd D.M. & Ellison N.B (2007). "Social Networking sites: definition, history and scholarship" Journal of Computer Mediated communication, 13(1) p.210-213

Colburn, Selene, & Haines, Laura (2012). Measuring libraries' use of YouTube as a promotional tool: An exploratory study and proposed best practices. Journal of Web Librarianship, 6(1), p.5–31.

Hupp, Jessica (2008). Useful Social Networking Tools for librarians. 8(6), p. 67. http://www.ilovelibraries.org/article/25-useful-social-networking-tools-librarians (accessed on 08/04/2016)

Kalpana S. S. and Pradip T P (2015). "Social networking tools for academic libraries". Knowledge librarian 2 (4) p. 5

Kroski, E (2007). Folksonomies and user-based tagging. In Courtney, N. (Ed), Library 2.0 and beyond: Innovative technologies and tomorrow's user. Westport, Conn: Libraries Unlimited. p.91-104.

Rathinasabapathy, G., and Mohana Sundari, T. (2005). Internet Using Pattern of Veterinary Science Students: A Case Study. In *Proceedings of the National Convention on Library and Information Networking (NACLIN). PES Institute of Technology* (pp. 358-365).

# 15

# Agriculture Institutional Repositories in India: Special Reference to Krishikosh

**Kundan Jha[1] Omprakash Sahu[2]**

[1] Assistant Librarian, Hon'ble Judges Library
High Court of Chhattisgarh, Bilaspur 495220, Chhatisgarh
*e-mail- kundanjha101@gmail.com*
[2] Computer Application Professional, Pali, District Korba, Chhatisgarh
*e-mail- om70228@gmail.com*

**ABSTRACT**

*This paper discuss about the Agriculture Institution Repositories (IRs) in India, its include merits, objectives, software requirements, growth and development. This paper also discusses about the KrishiKosh repository.*

***Key Words:*** *Institution Repository, KrishiKosh, National Agriculture Research System.*

## 1. Introduction

An Institution Repository support to provide research work in a digital platform to the use of information seeker i.e. researcher, faculty members, students and other learner persons, Institution Repository are digital archives where a universities or education institutions system research work is invented, accessible and preserved for future needs.

According to Clifford Lynch, "A university-based institution repository is a set of services that a university offers to the members of its community for the management and dissemination of digital materials created by the institution and its community members. It is most essentially an organizational commitment to the stewardship of these digital materials, including long-term preservation where appropriate, as well as organization and access or distribution"[1]

The creation and development of an institution repository system for any universities or research institutions is needed in the present scenario of information and communication technologies.

## 2. Objectives of Institutional Repository

The main objectives for having an educational institution repository are:

2.1. To create worldwide access for an institution's research work;

2.2. To collect content in a single platform;

2.3. To provide open access to educational institutions research work by self archiving;

2.4. To collect, store and preserve any other educational institute digital research assets.

## 3. Methodology

For data collection methodology, the main source of data collection are various observations which are from primary (journal and conference paper) and secondary source (Books and Handbook) of information and different type of search engine such us Met crawler, Google, yahoo, Msn, 123, and other type of search engine.

The data related to the research and development or educational institutions, institutional repositories have been accessed and retrieved from their respective institutions websites.

## 4. Software for Institutional Repositories

There are number of various types of Digital Library Software for the uses of Institutional Repositories are available i.e. 1. D Space (Digital Space), 2. E Print Archive, 3. Fedora (Flexible Extensible Digital Object and Repository Architecture), 4. GSDL (Green Stone Digital Library), 5. Ages Digital Library Software (My Ages), 6. AGES Software, 7. CDSware (CERN Document Server Software), 8. Dinest, 9. First Search, 10. Digi Tool, 11. Hyperion, 12. Genesha Digital Library Version 3.1, 13. LOCKSS (Lots of Copies Keep Stuff Safe), 14. CLOCKSS and 15. Meta Source. All the above software installation is very easy and stores various types of data like Ph. D theses, Dissertation, faculty publications, lecture notes, and other important project work and previous examination question papers.

## 5. Institutional Repositories in India

India has prospered through its strong academic and research institutions. The research and development institutions have also created and developed expertise in their specialized respective areas that are now recognized globally. Some of the leading institutions in India create, design and develop their own institution repositories which is listed in the Table 01(institutional repositories in India) are.

Table 01 *Institutional Repositories in India*

| S. No. | Name of the Repository and its / URL# | Host Institution | Place | Software Used | Subject Coverage |
|---|---|---|---|---|---|
| 01 | Eprints@Medknow Eprints. http://eprints.medknow.com | Medknow EPrints | Mumbai, Maharashtra | EPrints | Medical Science |
| 02 | OpenMED@NIC. http://openmed.nic.in | Bibliographic Information Division, National Informatics Centre (NIC) | New Delhi | EPrints | Health & Medicine |
| 03 | Indian Institute of Astrophysics Repository. http://prints.iiap.res.in | Indian Institute of Astrophysics | Bangaluru, Karnataka | DSpace | Physics & Astronomy |
| 04 | Digital Repository. http://dspace.rri.res.in | Raman Research Institute | Bangaluru, Karnataka | DSpace | Physics & Astronomy |
| 05 | DSpace @NCRA. http://ncralib1.ncra.tifr.res.in:8080/jspui | National Centre for Radio Astrophysics | Pune, Maharashtra | DSpace | Physics & Astronomy |
| 06 | Aryabhatta Research Institute of Observational Sciences (ARIES) Digital Repository. http://192.168.1.171:8080/jspuri | Aryabhatta Research Institute of Observational Sciences | Nainital, Uttarakhand | DSpace | Astronomy & Astrophysics |
| 07 | Library & Information Services. http://www.prl.res.in/~library | Physical Research Laboratory | Ahmedabad, Gujarat | GSDL | Physics, Astronomy, Astrophysics; Solar Physics & Planetary & Geosciences |
| 08 | DSpace@NCL. http://dspace.ncl.res.in | National Chemical Laboratory | Pune, Maharashtra | DSpace | Chemistry & Chemical Technology |
| 09 | EPrints@IISc. Open Access Repository of Indian Institute of Science Research Publications. http://eprints.iisc.ernet.in | Indian Institute of Science | Bangaluru, Karnataka | EPrints | Chemistry & Chemical Technology; Mathematics & Statistics; Astronomy & Astrophysics |

(Continued)

**Table 01 (Continued)**

| S. No. | Name of the Repository and its / URL# | Host Institution | Place | Software Used | Subject Coverage |
|---|---|---|---|---|---|
| 10 | Digital Knowledge Repository of Central Drug Research Institute (DKR@CDRI). http://dkr.cdri.res.in:8080/dspace/index.jsp | Central Drug Research Institute | Lucknow, Utter Pradesh | DSpace | Biology & Biochemistry; Health and Medicine |
| 11 | EPrints@SBT. http://epints.bicmku.in | School of Biotechnology, Madurai Kamraj University | Madurai, Tamil Nadu | EPrints | Biology & Biochemistry |
| 12 | EPrints@NII. http://eprints.nii.res.in | National Institute of Immunology | New Delhi | EPrints | Biology & Biochemistry |
| 13 | Open Access Agriculture Research Repository (OpenAgri). http://agropedia.iitk.ac.in/openaccess | Indian Institute of Technology | Kanpur, Uttar Pradesh | NA | Agriculture, Food & Veterinaries |
| 14 | EPrints@IARI. http://eprints.iari.res.in | Indian Agriculture Research Institute | New Delhi | EPrints | Agriculture, Food & Veterinaries |
| 15 | DSpace@IISR. http://220.227.138.214:8080/dspace/index.jsp | Indian Institute of Spices Research | Kozhikode, Kerela | DSpace | Botany, Horticulture, Biotechnology, Biochemistry |
| 16 | EPrints@CMFRI. http://eprints.cmfri.org.in | Central Marine Fisheries Research Institute | Kochi, Kerela | EPrints | Agriculture, Food & Veterinary; Biology; Environment & Ecology; Health & Medicine |
| 17 | Publication of the IAS Fellows. http://repository.ias.ac.in | Indian Institute of Science | Bangaluru, Karnataka | DSpace | Animal/Plant Science, Earth & Planetary Sciences; Mathematics & Statistics; Physics & Astronomy |
| 18 | National Aerospace Laboratories Institutional Repository (NAL) Repository. http://nal-ir.nal.res.in | Information Centre for Aerospace Science & Technology (ICAST) | Bangaluru, Karnataka | EPrints | Mathematics & Statistics; Technology General; Mechanical Engineering & Materials |

| S. No. | Name of the Repository and its / URL# | Host Institution | Place | Software Used | Subject Coverage |
|---|---|---|---|---|---|
| 19 | IMScEPrint Archive. http://www.imsc.res.in/eprints | Institute of Mathematical Science | Chennai, Tamil Nadu | EPrints | Mathematics & Statistics |
| 20 | EPrints@IIMT. http://eprints.iimt.res.in | Institute of Minerals & Materials Technology | Bhubaneswar, Orissa | EPrints | Material & Minerals Science |
| 21 | EPrints@IIITA. http://eprints.iiita.ac.in | Indian Institute of Information Technology | Allahabad, Uttar Pradesh | EPrints | Management; Science & Technology |
| 22 | EPrints@NIAS. http://59.90.235.217:8081 | National Institute of Advance Studies | Bangaluru, Karnataka | EPrints | Humanities, Natural Sciencs & Engineering; Social Sciences |
| 23 | EPrints@MoES. http://moeseprints.incois.gov.in/information.html | Ministry of Earth Science, Govt. of India | New Delhi | Eprints | Earth Science |
| 24 | eGyankosh. http://egyankosh.ac.in | Indira Gandhi National Open University | New Delhi | DSpace | Multidisciplinary |
| 25 | NISCAIR Online Periodicals Repository (NOPR). http://nopr.niscair.res.in | National Institute of Science Communication and Information Resources | New Delhi | DSpace | Multidisciplinary |
| 26 | ICRISAT Institutional Repository. http://openaccess.icrisat.org | International Crops Research Institute for the Semi Arid Tropics (ICRISAT) | Hyderabad, Telagana | DSpace | Agriculture |
| 27 | Shodhganga. http://shodhganga.inflibnet.ac.in | Information & Library Network Centre | Ahmadabad, Gujarat | DSpace | Multidisciplinary |
| 28 | Explorations – Open Access Repository (OAR) of Indian Theses. http://eprints.csirexplorations.com | CSIR Unit for Research & Development of Information Products | Pune, Maharashtra | EPrints | Multidisciplinary |

(Continued)

**Table 01 (Continued)**

| S. No. | Name of the Repository and its / URL# | Host Institution | Place | Software Used | Subject Coverage |
|---|---|---|---|---|---|
| 29 | E-Repository @IIHR (Knowledge Repository of) http://www.erepo.iihr.ernet.in | Indian Institute of Horticulture Research | Bangaluru, Karnataka | DSpace | Horticulture |
| 30 | KR@CIMAP. http://kr.cimap.res.in/index.jsp | Central Institute of Medicinal & Aromatic Plants | Lucknow, Uttar Pradesh | DSpace | Soil Science; Biology; Medical Science |
| 31 | Librarians' Digital Library. http://drtc.isibang.ac.in | Documentation Research & Training Centre (DRTC) & Indian Statistical Institute | Bangaluru, Karnataka | DSpace | Library & Information Science |
| 32 | Sarai Multimedia Digital Archive. http://archive.sarai.net/dspace | Sarai Multimedia Digital Archive | New Delhi | DSpace | Media Culture & Urban Spaces |
| 33 | DSpace@IIMK. http://dspace.iimk.ac.in | Indian Institute of Management | Kozhikode, Kerala | DSpace | Business, Management & Economics |
| 34 | EPeints@MDRF. http://mdrf-eprints.in | Madras Diabetes Research Foundation | Gopalapuram, Chennai | EPrints | Diabetes |
| 35 | DSpace@NCAOR. http://dspace.ncaor.org:8080 | National Centre for Antarctic Research | Goa | DSpace | Oceanography s |
| 36 | KrishiKosh http://krishikosh.egranth.ac.in | Indian National Agricultural Research System (NARS) | New Delhi | DSpace | Agriculture Science |
| 37 | KrishiPrabha http:// | Indian Council of Agriculture Research | New Delhi | DSpace | Agriculture Science |

# 6. KRISHIKOSH: An Institutional Repository under NARS

Theses and dissertations are known to be the rich and unique source of research information, often the only source of research work that does not find its way into various publication channels. Theses and dissertations remain an un-tapped and under-utilized asset, leading to unnecessary duplication and repetition that, in effect, is the anti-thesis of research and wastage of huge resources, both human and financial.

## 6.1. KRISHIKOSH: An Overview

"KrishiKosh" is the name coined to denote repository of agricultural science set-up by The Indian National Agricultural Research System (NARS). The word "Krishi" means a subject related with agricultural activities. The "Kosh" means collection of information. KrishiKosh stands for the reservoir of Indian agricultural intellectual output stored in a repository hosted and maintained by the National Agricultural Innovation Project (NAIP).

The KrishiKosh is an Institutional Repository developed under the under National Agricultural Research System (NARS). In this project, the intellectual output of Indian NARS in various document forms is captured, preserved/ archived to enable users to access the content online. It is a unique repository of Knowledge in agriculture and allied sciences, having a collection of theses, rare and valuable books, institutional publications, technical bulletins, project reports, lectures, preprints, reprints, field records and other documents available in different libraries of Research Institutes and State Agricultural Universities (SAUs) spread all over the country. KrishiKosh acts as a digital platform to preserve the institution's intellectual assets and to also manage its use through open access mandate.

"KrishiKosh" support various document formats for submission of agricultural resource in the KrishiKosh. Portable document format (PDF) is the perfect document format in this repository.

## 6.2. Features of KrishiKosh

The features of KrishiKosh are:

6.2.1 Improve Accessibility

6.2.2 Enhanced Search Ability

6.2.3 Preservation

6.2.4 Content Selection

## 6.3. Communities in KrishiKosh

**Table 02 Communities in KrishKosh**

| S. No. | Universities/ Institute/ Research Centre | Contact Information |
|---|---|---|
| 01 | Anand Agricultural University | http://www.aau.in/ Anand Agricultural University, Anand388110 Gujarat (INDIA,.) . |
| 02 | Bihar Agricultural University | http://www.bausabour.ac.in/<br>Bihar Agricultural University, Sabour Bhagalpur, Bihar India- 813210 |
| 03 | Central Arid Zone Research Institute | http://www.cazri.res.in/<br>Director Central Arid Zone Research Institute Near Industrial Training Institute (ITI) Light Industrial Area Jodhpur - 342 003 (Rajasthan) - INDIA |
| 04 | Central Inland Fisheries Research Institute | http://www.cifri.ernet.in/<br>The Director, ICAR - Central Inland Fisheries Research Institute, Monirampur (Post), Barrackpore Kolkata, West Bengal - 700 120 |
| 05 | Central Institute for Research on Cotton Technology | http://circot.res.in/circot/<br>**Central Institute for Research on Cotton Technology,** Adenwala Road, Matunga(East), Mumbai-400 019. |
| 06 | Central Institute Of Fisheries Education | www.**cife**.edu.in<br>Panch Marg,Off Yari Road, Versova, Andheri West, Mumbai, Maharashtra 400061 |
| 07 | Central Institute of Fisheries Technology | http://www.cift.res.in/<br>Central Institute of Fisheries Technology CIFT Junction, Willingdon Island Matsyapuri P.O., Cochin-682 029, Kerala |
| 08 | Central Institute of Freshwater Aquaculture | http://www.cifa.in/ **irector**<br>**Central Institute of Freshwater Aquaculture** Kausalyaganga, Bhubaneswar-751002, ODISHA, INDIA |
| 09 | Central Marine Fisheries Research Institute | http://www.cmfri.org.in/<br>Central Marine Fisheries Research Institute, Abraham Madamakkal Rd, Ayyappankavu, Ernakulam, Kerala 682018 |

| S. No. | Universities/ Institute/ Research Centre | Contact Information |
|---|---|---|
| 10 | Central Plantation Crops Research Institute | http://www.cpcri.gov.in/<br>Central Plantation Crops Research Institute, Kasaragod, Kerala, India |
| 11 | Central Potato Research Institute | http://cpri.ernet.in/<br>**Central Potato Research Institute** Shimla-171001,HP, INDIA |
| 12 | Central Rice Research Institute | http://www.crri.nic.in/<br>**Director** ICAR-National Rice Research Institute Cuttack (Odisha) 753 006, India |
| 13 | Central Soil Salinity Research Institute | http://www.cssri.org/<br>**CSSRI Head Office:** Zarifa Farm, Kachawa Road, Karnal - 132001 Haryana (India) |
| 14 | Chaudhary Charan Singh Haryana Agricultural University | Hisar, Haryana 125004 |
| 15 | Chhattisgarh Kamdhenu Vishwavidyalaya | www.cgkv.ac.in<br>Camp Office, Near 36 Mall, NH6, Raipur, Chhattisgarh 492001 |
| 16 | Directorate of Oilseeds Research | http://icar-iior.org.in/<br>Rajendranagar mandal, Hyderabad, Telangana 500030 |
| 17 | Govind Ballabh Pant University Of Agriculture And Technology | http://www.gbpuat.ac.in/**Error! Hyperlink reference not valid.**<br>**Pantnagar - 263145, District - Udham Singh Nagar, Uttarakhand (India)** |
| 18 | Indian Agricultural Research Institute | http://www.iari.res.in/**Error! Hyperlink reference not valid.**<br>Indian Agricultural Research Institute, New Delhi |
| 19 | Indian Agricultural Statistics Research Institute | http://iasri.res.in/**Error! Hyperlink reference not valid.**<br>Library Ave, Pusa, New Delhi, Delhi 110012 |
| 20 | Indian Council of Agricultural Research | http://www.icar.org.in/**Error! Hyperlink reference not valid.**<br>**Indian Council of Agricultural Research,** Krishi Bhavan, New Delhi 110 001 |
| 21 | Indian Grassland and Fodder Research Institute | http://www.igfri.res.in/**Error! Hyperlink reference not valid.**<br>Near Pahuj Dam, Gwalior Road, Jhansi - 284 003 (UP) India |

(Continued)

**Table 02 (Continued)**

| S. No. | Universities/ Institute/ Research Centre | Contact Information |
|---|---|---|
| 22 | Indian Institute of Horticultural Research | http://www.iihr.res.in/**Error! Hyperlink reference not valid.**<br>Director<br>ICAR-IIHR, Hessaraghatta lake post,<br>Bengaluru-560 089. |
| 23 | Indian Institute of Spices Research | http://www.spices.res.in/**Error! Hyperlink reference not valid.**<br>Chelavoor, Kozhikode, Kerala State 673012 |
| 24 | Indian Veterinary Research Institute, Izatnagar | http://ivri.nic.in/**Error! Hyperlink reference not valid.**<br>Izatnagar, Bareilly, Uttar Pradesh 243122 |
| 25 | Indian Veterinary Research Institute-Mukteswar | http://www.ivri.nic.in/campuses/mukt/default.aspx**Error! Hyperlink reference not valid.**<br>Mukteshwar, Uttarakhand 263138 |
| 26 | Indira Gandhi Krishi Vishwavidyalaya, Raipur | http://igau.edu.in/**Error! Hyperlink reference not valid.**<br>Indira Gandhi Agricultural University, Krishak Nagar, 492012, Raipur, CG, India |
| 27 | Mahatma Phule Krishi Vidyapeeth | http://www.mpkv.ac.in/**Error! Hyperlink reference not valid.**<br>Ahmednagar, Rahuri, Maharashtra 413722 |
| 28 | National Academy Of Agricultural Research Management | http://www.naarm.ernet.in/index.php?lang=en**Error! Hyperlink reference not valid.**<br>ICAR - National Academy of Agricultural Research Management, Rajendranagar, Hyderabad 500030, Telangana, India |
| 29 | National Bureau Of Agriculturally Important Insects | http://www.nbair.res.in/**Error! Hyperlink reference not valid.**<br>The DirectorNational Bureau of Agricultural Insect ResourcesP.Bag No:2491, H.A. Farm PostBellary Road, Bengaluru - 560 024.Karnataka, INDIA. |
| 30 | National Bureau Of Plant Genetic Resources | http://www.nbpgr.ernet.in/**Error! Hyperlink reference not valid.**<br>**National Bureau of Plant Genetic Resources**, Indian Council of Agricultural Research, Ministry of Agriculture (Govt. of India), Pusa Campus, New Delhi-110012, INDIA |

| S. No. | Universities/ Institute/ Research Centre | Contact Information |
|---|---|---|
| 31 | National Bureau Of Soil Survey And Land Use Planning | http://www.nbsslup.in/**Error! Hyperlink reference not valid.**<br>NH6, Vayusena Nagar, Nagpur, Maharashtra 440007 |
| 32 | National Dairy Research Institute | http://www.ndri.res.in/ndri/Design/Index.html**Error! Hyperlink reference not valid.**<br>**National Dairy Research Institute,** Indian Council of Agricultural Research, Ministry of Agriculture (Govt. of India), Pusa Campus, New Delhi-110012, INDIA |
| 33 | National Institute of Research on Jute and Allied Fibre Technology | http://www.nirjaft.res.in/**Error! Hyperlink reference not valid.**<br>12, Regent Park, Kolkata, West Bengal 700040 |
| 34 | National Research Centre on Camel, Bikaner | http://www.nrccamel.res.in/**Error! Hyperlink reference not valid.**<br>Post Box - 07, Jodhpur Bypass, Rajasthan 334001 |
| 35 | National Research Centre On Coldwater Fisheries - Bhimtal | http://www.dcfr.res.in/**Error! Hyperlink reference not valid.**<br>Anusandhan Bhawan, Industrial Area, Bhimtal - 263136, Distt: Nainital, Uttarakhand, India |
| 36 | Navsari Agricultural University | http://www.nau.in/**Error! Hyperlink reference not valid.**<br>Dandi Rd, Navsari, Gujarat 396450 |
| 37 | Orissa University of Agriculture and Technology | http://www.ouat.ac.in/**Error! Hyperlink reference not valid.**<br>Orissa University of Agriculture & Technology Bhubaneswar-751003 Orissa, India. |
| 38 | Professor Jayashankar Telangana State Agricultural University | http://www.pjtsau.ac.in/**Error! Hyperlink reference not valid.**<br>**Professor Jayashankar Telangana State Agricultural University (PJTSAU)**<br>Rajendranagar, Hyderabad-500 030 Telangana, India. |
| 39 | Project Directorate on Poultry | http://www.pdonpoultry.org/pdpnew/**Error! Hyperlink reference not valid.**<br>DIRECTORATE OF POULTRY RESEARCH, Rajendranagar, Hyderabad 500 030, Telangana, India |
| 40 | Sher-e-Kashmir University of Agricultural Sciences and Technology-Jammu | http://www.skuast.org/**Error! Hyperlink reference not valid.**<br>Chatha, Jammu, Jammu and Kashmir 180009 |
| 41 | Tamilnadu Agricultural University | ***http://www.tnau.ac.in***<br>Marudhamalai Rd, Coimbatore, Tamil Nadu 641003 |

(Continued)

**Table 02 (Continued)**

| S. No. | Universities/ Institute/ Research Centre | Contact Information |
|---|---|---|
| 42 | Tamil Nadu Veterinary and Animal Sciences University | http://www.tanuvas.tn.nic.in/<br>Madhavaram Milk Colony, Chennai, Tamil Nadu 600051 |
| 43 | University of Agricultural & Horticultural Sciences, Shivamogga | http://uahs.in/Error! Hyperlink reference not valid.<br>SH 57, Navule, Shivamogga, Karnataka 577216 |
| 44 | University of Agricultural Sciences, Bengaluru | http://www.uasbangalore.edu.in/Error! Hyperlink reference not valid.<br>NH7, near L&T Factory,Gandhi Krishi Vignana Kendra, Bengaluru, Karnataka 560065 |
| 45 | University of Horticultural Sciences, Bagalkot | http://www.uhsbagalkot.edu.in/Error! Hyperlink reference not valid.<br>University of Horticultural Sciences, Udyanagiri, Near Seemekeri Cross, Hubli Bypass Road, Navanagar, Bagalkot-587104 Karnataka, India. |

## 6.4. Categories of Accessibility

All of NARS's holdings are classified under the following three access categories:

**6.4.1 Public Access:** Any Record that can be made available to public at large shall fall under this category

**6.4.2 Privileged Access:** Records classified under this category shall be accessible to only to those individuals or organizations that have a privileged status with NARS (such as other national / state archives / research and academic institutes / eminent researchers etc.). Others (the world at large) would have to seek prior permission / approval from IARI to access any Record classified as Privileged Access.

**6.4.3 Prohibited Access:** Records which are accessible ONLY to NARS authorized officials, due to their confidential and sensitive nature as defined by statutory rules and regulation.

## 7. Conclusion

Institution Repositories have great powerful potential for providing visibility and impact of institution research work. Institutional repositories enhance learning, research activity and are considered as a boon to the scholarly communities in any country. Especially, the KrishiKosh is playing a vital role in disseminating agricultural knowledge among stakeholders of agricultural and related disciplines in India. All Universities should come forward to establish their own institutional repositories and upload their publications to ensure wider access across the globe.

## References


Consortium for e-Resources in Agriculture, http://cera.jccc.in (Accessed 07/05/2016)

Gohain, R. R. (2011, September). Current Trend and Development of Institutional Repositories in India. International Journal of Information Research, I (1), 01-19.

Gopikuttan, A. (2008). Current Developments in Library & Information Services. Journal of Information Science and Technology, I (1), 89-100.

http://shodhganga.inflibnet.ac.in (Accessed dated 09/04/2016)

Indian Council of Agricultural Research, http://www.icar.org.in (Accessed 07/05/2016)

Jain, A. K., & Veeranjaneyulu, K. (2013). Retrieved from https://www.google.co.in/url?sa=t&rct=j&q=&esrc=s&source=web&cd=1&cad=rja&uact=8&ved=0ahUKEwizlbCCvMrMAhWKHI4KHQv4D3UQFggfMAA&url=http%3A%2F%2Foasis.col.org%2Fbitstream%2Fhandle%2F11599%2F1830%2F2013_Jain%2526Veeranjaneyulu_NARES.pdf%3Fsequence%3D1%26isAl (Accessed 07/05/2016)

Kamila, K. (2009). Institution Repository Project in India. 7th International CALIBER-2009 (pp. 128-132). Puducherry: INFLIBNET Centre, Ahmadbad.

KrishiKosh - a Digital Repository of NARES http://krishikosh.egranth.ac.in (Accessed 07/05/2016)

Lynch, C. (2003). Institution Repositories: Essential Infrastructure for Scholarship in the Digital Age. Retrieved May 04, 2016, from ARL Bimonthly Report: www.arl.org/newsltr/226/ir.html

National Agriculture Innovation Project, http://www.naip.icar.org.in (Accessed 07/05/2016)

Strengthening of digital library and Information Management under NARES (eGranth), www.egranth.ac.in (Accessed 07/05/2016)

Veeranjaneyulu, K. (n.d.). Retrieved from https://www.google.co.in/url?sa=t&rct=j&q=&esrc=s&source=web&cd=2&cad=rja&uact=8&ved=0ahUKEwizlbCCvMrMAhWKHI4KHQv4D3UQFgglMAE&url=http%3A%2F%2Flib.hku.hk%2Fetd2013%2Fpresentation%2FVeeranjaneyulu%2520-%2520Krishikosh.pdf&usg=AFQjCNERURDae5oy7WmzfG32MmKq4p (Accessed 07/05/2016) AgriCat - a Group catalogue of Agricultural Libraries of NARES, www.agricat.worldcat.org (Accessed 07/05/2016)

# 16

# User's Behavior on Usage of e-Resources in an Indian University: A Case Study

**Manoj Mishra[1], D B Ramesh[2] and Rabindra K Mahapatra[3]**

[1] Assistant Librarian, S 'O' A University, Bhubaneswar, Odisha
*e-mail:* mishra_manoj75@rediffmail.com
[2] Chief Librarian, S 'O' A University, Bhubaneswar, Odisha
*e-mail:* dbramesh@soauniversity.ac.in
Assoc. Prof., Central University, Tripura,
*e-mail: mahapatrark_1962@rediffmail.com*

**ABSTRACT**

*E-resources are the important collections of the libraries in this digital era. E-resources are being browsed by the help of ICT applications in the libraries which needs a comprehensive task to run successfully to meet the user's demands instantaneously. These resources are very useful for the faculties and research scholars of various institutions for better output. Electronic resources can be accessed now-a-days through desktop, Laptop, mobiles and i-phone etc. Electronic information sources are taking a major role in part of the academic institutions. The purpose of this study is to find out the level of awareness and use of electronic resources among the users of the Central Library, The library of Siksha 'O' Anusandhan University is well equipped with the electronic resources and determined to serve the users to improve access, update and evaluate academic and research information. The library is subscribing e-resources from eleven publishers and aggregators to access e-resources for the patron users of Library: The data collected for the study on the usage of e-resources from the Library usage statistics during last year and presented with suitable figures and tables with findings.*

## 1. Introduction

E-resources are the important resources of the libraries in this current situation. E-resources are being browsed by the help of Information Communication Technology applications in the libraries. Library professionals and authorities should have keen interest towards the application of ICT in information centers for better utilization of electronic resources. These resources are very useful for the faculties and research scholars of various institutions for better output. The examination of the existence of different databases in Guru Govind Singh Indraprastha University library emphasizes the importance and as well as the preference of online resources among the faculties and research scholars. Usage of computers in present situation has a strong demand in the radical changes of library services. In present century the computers have a great role in collecting, storing, accessing, organizing, retrieving and utilizing the e-resources available in the society. Libraries have seen more transfiguration in this electronic era both in their collection development and in their different modules of library services. The prime concept of this metamorphosis is that print environment is increasingly paving the way to digital structure of the resources.

There are more relevant studies on usage of electronic resources by faculties, research scholars and under graduate and post graduate students of universities and the various research organizations. 78 % of the interviewers show their interest in using the UGC–Infonet e-journals and agreed that the research work depends on these journals where they need current article alert services and electronic document supply services (Madhusudan 2008). The importance of e-journals is reflected as almost sixty eight percent of research scholars of faculty of science and sixty nine percent of research scholars of engineering are acquainted with e-journals and 35 % of science students update their knowledge with the use of e-journals and 24 % of engineering students use these journals for their study purpose. Kaur and Verma tell that the users need the recent and modern techniques available like audio-visual source, CD and DVD, online database and the other web resources. It is understood that the users are interested to utilize the e-resources frequently because of it's speed of availability and easy to access. There is a bare need of training for the users to motivate them towards using of the e-journals. Kennedy proposes to include the library catalogue in library URL to make some solution in the maintenance of increased website links.

The academic research scholars are more dependent on electronic journals in present digital environment. Navjyoti finds that the expeditious publication and connection on the internet are the basic factors for which the electronic resource users are attracted. Further more studies conducted on ICT application in libraries by different authors from various regions and the outlook provides a healthy platform from which to clinch the status of ICT applications in the libraries of developing countries. Hopkinson, Seneviratne and Amaraweera, Singh and Vasishta studied on the application of ICT in developing countries and highlighted the status in developed countries. Print format is gradually paving the way to the electronic form of library resources. Madhusudan stated in his study that more users admit that using of e-journals of the UGC-Infonet creates high potentiality of the research work and also they require the current article alert service with the electronic document delivery service.

There was a time; the librarians in Pakistan were reluctant for library automation. They feel that introducing the automation, professionalism can be spoiled. When they got through the advantage of this service, they apply their knowledge to take this service into action. ICT applications in libraries need a comprehensive task to run successfully to meet the users demand. It will be unsuccessful if the users face problem while accessing the e-resources, if they face the non-co-operation from the library authorities and insufficient budget allocation for this purpose.

## 2. Need and Advantages of e-Resources

E-resources are basically useful for the academic and research purpose. Therefore, the collection building activity of the libraries is influenced by the e-resources. The examples of various e-resources are as follows:

- E- Books
- E - Images
- E- Journals
- E- Newsletters
- E- Newspapers
- E- Thesis
- E-Audio-Visual
- E-Databases

The importance of e-resources is gradually increasing in libraries due to the user expectation and easy to access. The advantages of e-resources enforce the library authorities to add these resources in library subscription to facilitate their service to the end user. The working librarians should meet the challenge of finding solutions to manage the electronic resources from selection to licensing. The advantages of e-resources in comparison to the print resources cannot be ignored. The advantages of e-Resources are as follows:

- One can access more number of literatures at a time in e-Resource than print resource.
- The user can gather more than 10 years of back volumes of any e-Resource subscription which is a hard task to collect in print resource.
- There is the facility for the users to search on their required field from many publications without consuming more time.
- As these are electronically available, it is very fast to obtain current issues to update knowledge.
- E-resource provides accurate and current materials than any search engine available for access.
- E-resources are free to access in the libraries to any kind of user during the open hours of the concerned libraries using the library cards.

The purpose of the study is to find out the level of awareness and use of electronic resources among the users of the Central Library. The data collected from the Library statistics for last year (January to December) were scrutinized, prepared to analyze with the help of statistical tools and findings were drawn. The faculty members in various positions are serving in the institute need information for better teaching learning process. The Library is a constituent wing of this University. It has open role on outstanding research and quality education. This happens to be the liveliest place with Wi-Fi providing a safe, comfortable and friendly environment that helps learning and improvement of knowledge and promotes discovery and scholarship. The important role of the library is to facilitate the creation of new knowledge through acquisition, preservation and dissemination of knowledge resources and providing for value added services through proper organization.

Besides, the library is well equipped with advanced technical facilities like e-learning, access to internet and web resources including on-line journals and e-books. The entire housekeeping work of the library is computerized. This library has a good collection of both text and reference books. For better teaching and research process, the library subscribes hard copy journals of both Indian & foreign. Besides that it has the facility to access current on-line journals from publisher and aggregators on different subjects: Engineering, Medicine, Management, Agriculture, Nursing, Law etc.ne. The library has also a good collection of CDs & DVDs of different sectors and valuable theses and dissertations for reference purpose.

## 3. E-Resources at Siksha 'O' Anusandhan University, Bhubaneswar, Odisha

There is a comfort access to the subscribed e-resources of this University. Library authorizes the users and the sites for better utilization of the resources for research development and fruitful teaching. The walk-in-use process is also available to browse the authentic information from the sources. All the users can access their required information from anywhere in the University campus without any hesitation with the registered IP and proxy server. This facility is commonly known as remote use. List of e-resources subscribed to the University Library:

- *SCOPUS -http://www.scopus.com (Elsevier) Fulltext Databases*
- *IEEE Xplore Digital Library - http://ieeexplore.ieee.org*
- *ScienceDirect - http://www.scincedirect.com*
- *EBSCO - http://search.ebscohost.com*
- *PROQUEST - http://search.proquest.com*
- *RSC - http://pubs.rsc.org*
- *IOP - http://iopscience.iop.org*
- *Taylor & Francis (http://www.tandfonline.com*

- *Annual Reviews - http://www.annualreviews.org*
- *ProQuest's Ebrary - http://site.ebrary.com/lib/soauniversity*
- *Ebscohost's ECM - http://ecm.ebscohost.com*

## 4. Results

The data collected from the Library statistics for last year (January to December) were scrutinized, prepared to analyze with the help of statistical tools and findings were drawn. It is found from this study that averagely 1000 users visit library every day to meet their satisfaction. They use the library resources and collect their information to enhance their knowledge (Fig.- 1).

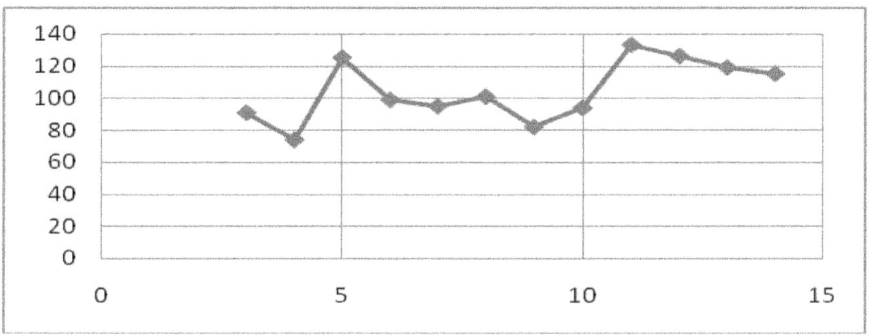


**Fig.– 1: Average no. of Users visit library in a Day**

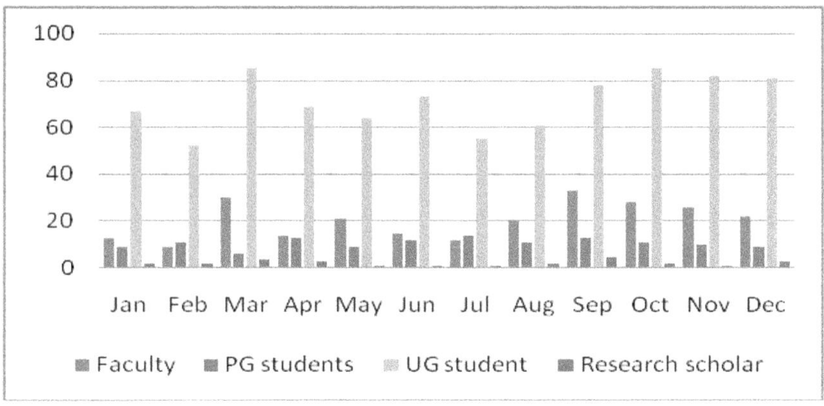


**Fig.– 2: Distribution of Library Users**

Although more numbers of under graduate students visit library every day, still faculty members with post graduate students and research scholars are attracted to the library for utilization of library resources (Fig.- 2).

Averagely 20 % of faculty, 10 % of PG students, 68 % of UG students and 2 % of research scholar visit the library among the total users of a single day approximately

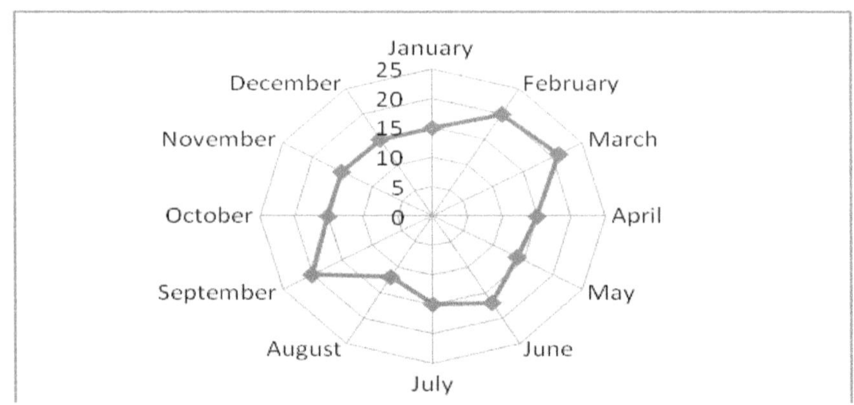


**Fig.– 3: Average no. of users visit e-library**

The faculties and research scholars mainly use the e-resources for research purpose and for better teaching process with update knowledge. This survey finds that approximately 750 users use the e-resources subscribed to this library daily (Fig.- 3).

***Table – 1: Use of*** **Electronic Databases (Bibliographic & Full Text)**

| Electronic Databases (Bibliographic & Full Text) | USERS | | | |
|---|---|---|---|---|
| | Faculty | Research Scholars | PG students | UG students |
| SCOPUS (Elsevier) Full text Databases | 84% | 93% | 68% | 7% |
| IEEE Xplore Digital Library | 56% | 73% | 48% | Rare |
| ScienceDirect | 79% | 99% | 94% | 3% |
| EBSCO | 84% | 78% | 89% | 2% |
| PROQUEST | 73% | 69% | 57% | 4% |
| RSC | 32% | 48% | 15% | Rare |
| IOP | 43% | 87% | 25% | Rare |
| Taylor & Francis | 15% | 47% | 32% | Rare |
| Annual Reviews | 12% | 26% | 17% | Rare |

All the users use full text database and bibliographic database for their purpose which is relevant to their study, research and teaching. (table - 1)

**Table - 2: Use of Electronic Books / Databases**

| E-Book Databases | USERS | | | |
|---|---|---|---|---|
| | Faculty | Research Scholars | PG students | UG students |
| ProQuest's Ebrary | 48% | 52% | 84% | 79% |
| Ebscohost's ECM | 35% | 49% | 73% | 67% |

All the users are familiar with the use of e-books. 48 % of faculty members, 52 % of research scholars, 84 % of post graduate students and 79 % of under graduate students use Ebrary from Proquest database where 35 % of faculty members, 49 % of research scholars, 73 % of post graduate students and 67 % of under graduate students use ECM of EBSCO database. (table - 2)

## 5. Findings

Major findings of the study are as follows:

- An average of 1000 users visits the library every day.
- Although more numbers of under graduate students visit library every day, still faculty members with post graduate students and research scholars also visit library for utilization of library resources.
- Among the daily visitors, an average 20 % of faculty, 10 % of PG students, 68 % of UG students and 2 % of research scholar visit the library.
- The faculties and research scholars mainly use the e-resources for research and teaching – learning system.
- Approximately, 1500 users daily visit E-Library to use the subscribed e-resources
- They prefer to have the full text of the paper than the abstract and use ScienceDirect, ProQuest and Ebsco databases subscribed by the University.
- They search bibliographic databases such as PubMed, Agricola and Scopus for the literature survey of their topics of interest.
- All the users are familiar with the use of e-books such as E-brary. 48 % of faculty members, 52 % of research scholars, 84 % of post graduate students and 79 % of under graduate students use Ebrary in different times.
- 35 % of faculty members, 49 % of research scholars, 73 % of post graduate students and 67 % of under graduate students use E-books of ECM provided by EBSCO database.
- There is a need of infrastructure and training program for the users for better browsing of the electronic resources.
- Usage of electronic materials is a common process among the users of IMS.

## 6. Conclusion

Average number of users visit to the library every day quite satisfactory. Although more numbers of under graduate students visit library every day, still faculty members with post graduate students and research scholars also visit library for utilization of library resources. The faculties and research scholars mainly use the e-resources for research, patient care and teaching. Preference is more to have the full text of the paper than the abstract and use ScienceDirect, ProQuest and Ebsco databases subscribed by the University. They search for bibliographic databases like PubMed and Scopus for the literature survey of their topics of interest. All the users are familiar with the use of e-books E-brary, ECM provided by EBSCO database. There is a need of infrastructure and training programme for the users for better browsing of the electronic resources. Usage of electronic materials is a common process among the users. The electronic information sources are taking a major role in part of the academic institutions. The libraries which need a comprehensive task to run successfully to meet the user's demands instantaneously. The electronic materials are very useful for the faculties and full time researchers of academic institutions for better output.

## References

Eqbal Monawwer and Khan Azhar Shah (2007). Use of Electronic Journals by the Research Scholars of Faculty of Science and Faculty of Engineering. In:NACLIN- 2007, pp309–319.

Higher Education Commission. Digital library: a programme of Higher Education Commission [Internet] the Commission; 22 Aug 2011 [cited 22 Oct 2014]. http://www.digitallibrary.edu.pk

Hopkinson A (2009). Library automation in developing countries: the last 25 years. *Inf. Dev.,* 25(4), 304–12.

Kaur Baljinder&Verma Rama (2006). Use of Electronic Resources at TIET Library Patiala: A Case Study. *ILA Bulletin*, 42( 3), 18–20.

Kennedy, P (2004). Dynamic Web pages and the library catalogue.*The Electronic Library*, 22(6), 480-486.

Madhusudan M. Use of UGC infonet e-journals by research scholars and students of University of Delhi, Delhi. *Library Hi Tech*, 26(3), 369–386.

Mairaj, Muhammad Ijaz (2012). Application of information and communication technologies in libraries in Pakistan. *J Med Libr Assoc.*, 100 (3), 218 – 222.

Malik K.M. (1996) Status of library automation in Pakistan. *Lib Rev.*, 45(6):36–42.

Naidu GHS, Rajput Prabhat&MotiyaniKavita (2007). Use of Electronic Resources and Services in University Libraries: A Study of DAVV Central Library, Indore. In *NACLIN -2007*, pp309–319.

Narayana Poornima&Goudar I.R.N (2005). E-Resources Management through Portal: A Case Study of Technical Information Center. In International Conference on Knowledge Management (ICIM2005), 22-25 Feb. 2005, pp1–19.

Navjyoti, A. (2007). Snapshot of E-Journals' Adopters (Research Scholars) of Guru Nanak Dev University. In *NACLIN - 2007*, pp432–442.

Ramzan M &Singh D (2009). Status of information technology applications in Pakistani libraries.*Electron Lib.*, 27(4), 573–87.

Seneviratne G.P &Amaraweera A.J (2002). Automation of library operations in Sri Lanka: a cost effective web-based solution. Inf. Dev., 18(2), 111–15.

Sharma, Chelan (2009). Use and Impact of e-Resources at Guru Govind Singh Indraprastha University (India): A Case Study: *Electronic Journal of Academic and Special Librarianship,* 10 (1),1 – 8.

Sharma, Chetan (2009). Use and Impact of E-Resources at Guru Govind Singh Indraprastha University (India): A Case Study. *Electronic Journal of Academic and Special Librarianship,* 10(1),

Singh A (2003).. Library automation and networking software in India: an overview. Inf. Dev.,19(1), 51–5.

Vasishta S (2008) . Modernization of library and information services in technical higher education institutions in North India: state-of-the-art report. *IFLA J.* 34, pp286–94.

# 17

# Role of LIS Professionals in Knowledge Management

**Dr.S. Jahan Ara[1] and Dr. D. Chandran[2]**

[1] Lecturer in Library and Information Science, SJGC Degree College Kurnool- 518 001, Andra Pradesh. *e-mail: jahanaramqbl@gmail.com*

[2] Former Professor, Dept. of Library and Information Science Sri Venkateswara University, Tirupati – 517 501, Andra Pradesh *e-mail: chandrand2001@gmail.com*

**ABSTRACT**

*This paper attempts to introduce the latest trends in the Library and Information dissemination services for Knowledge Management and study of the Knowledge Management Components for the progress of the Organization and explains about practicing and Managing the Knowledge in Indian context which suits our organizational set up.*

***Keywords:*** *Knowledge Management, Library and Information Science, India*

## 1. Introduction

Libraries have an important role to play in social, cultural, political and economic development of a country. Libraries have a noble mission of imparting education and knowledge to the faculty, research scholars, students and other patrons. The present generation of library professionals recognizes that the advent of information revolution in the field of science and technology made their role to play in improving the quality of services and also to become tech-savvy.

## 2. Role of Libraries in Knowledge Management:

According to S.R. Ranganathan there are four categories of users namely, freshman, general readers, ordinary inquirer and the specialist inquirer. The advent of Internet facilities and availability of electronic or digital libraries overshadow the existence of traditional or conventional libraries. We assume that these technical advances change the shape of libraries and their services. Library automation surely help in monitoring the quick access of information to the users not in the institution but also in the campus whether they sit at the café, canteen or hostel irrespective of the time and place.

## 3. Need of the study

Information technology and computer based systems are now in domination and revolutinionizing in every field of life. The change is clearly seen in the field of Library and information science also. Digital libraries have brought the concept of "global libraries". So, Knowledge management processes are essentially centered round knowledge workers and an organization can use this knowledge as a strategic resource to attract, retain and motivate them in turn the development and progress in the Organization takes place.

## 4. Objectives of the study

1. Introducing the latest trends in the Library and Information dissemination services for Knowledge Management.
2. Study of the Knowledge Management Components for the progress of the Organization.
3. Practicing and Managing the Knowledge in Indian context which suits our organizational set up.

## 5. Defining Knowledge and Knowledge Management (KM)

Knowledge and Information are two different things. Information is facts and data that are organized to describe a situation. Knowledge is something that covers the truths and beliefs, perspectives and concepts, judgments and expectations and knows how.

Knowledge Management is a systematic attempt to use knowledge within an organization to improve overall performance.

"Knowledge management involves efficiently connecting those who know with those who need to know and converting personal knowledge into organizational knowledge".

Knowledge Management (KM) refers to a multi-disciplined approach to achieving organizational objectives by making the best use of knowledge. KM focuses on processes such as acquiring, creating and sharing knowledge and the cultural and technical foundations that support them.

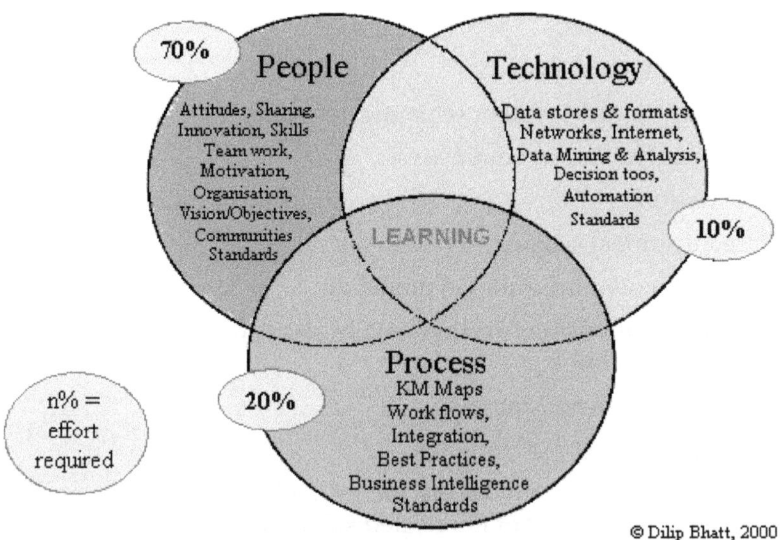


## 6. Knowledge Management may be viewed in terms of:

- People – how do you increase the ability of an individual in the organization to influence others with their knowledge
- Processes – Its approach varies from organization to organization. There is no limit on the number of processes
- Technology – It needs to be chosen only after all the requirements of a knowledge management initiative have been established.

Or

- Culture –The biggest enabler of successful knowledge-driven organizations is the establishment of a knowledge-focused culture
- Structure – the business processes and organizational structures that facilitate knowledge sharing
- Technology – a crucial enabler rather than the solution.

## 7. Knowledge Management Related To:

Knowledge management draws from a wide range of disciplines and technologies:

- Cognitive science
- Expert systems, artificial intelligence and knowledge base management systems (KBMS)

- Computer-supported collaborative work (groupware)
- Library and information science
- Technical writing
- Document management
- Decision support system semantic networks
- Relational and object databases
- Simulation
- Organizational science
- object-oriented information modeling
- electronic publishing technology, hypertext, and the World Wide Web; help-desk technology
- full-text search and retrieval
- performance support systems

Although around 20 kinds of disciplines and study areas were listed above, there is no way to include all of the related subjects to knowledge management.

## 8. The Value of Knowledge Management in practicing and monitoring in Library services

Knowledge Management in Library services will pave the way to:

- Foster innovation by encouraging the free flow of ideas
- Improve decision making
- Improve customer service by streamlining response time
- Boost revenues by getting products and services to market faster
- Enhance employee retention rates by recognizing the value of employees' knowledge and rewarding them for it
- Streamline operations and reduce costs by eliminating redundant or unnecessary processes

## 9. Knowledge Management Today by the LIS professionals

Implementation of Knowledge Management in College and University libraries is a complex process which needs strategic approach to achieve the mission and vision of the College and university. Based on a clear understanding of needs and priorities the Librarian and a team of the teaching staff along with students, researchers, staff should develop a realistic plan and assign resources for implementation process. The following are the steps included in the process:

| Analyze the existing Infra structure |
|---|
| Align Knowledge Management and Innovation strategy |
| Design KM Architecture. |
| Design KM team |
| Unite people, process and technology |
| Develop the KM system |
| Performance Evaluation. |

## 10. Benefits of Knowledge Management in the Colleges

1. Facilitate the conversion of Implementation in to knowledge.
2. It is very essential to codify tacit knowledge into explicit knowledge.
3. Ready access to relevant knowledge for key professionals and decision makers.
4. Minimization of duplication of efforts.
5. Improved innovation and new product development,
6. Minimizing the impact of loss of experienced knowledge workers.

## 11. Role of Librarians /Libraries as "Knowledge Houses to Manage" the Information:

Libraries are the Information resource Centers. They deal with both the categories of knowledge, tacit and explicit.

Tacit knowledge for the library personnel and explicit knowledge for the end user. Library personnel must have clear view of the "know-how" of information sources, management, retrieval and dissemination as well as global access to information.

The tacit knowledge helps the end users to gain explicit knowledge. They must be guided to the gateway of knowledge. The objective of knowledge management in libraries is to promote knowledge innovation, relationship in and between libraries and between library and user.

## 12. Suggestion

Knowledge Commission report strengthen the necessity of the Education in every field and the Libraries have the important role to play in developing the country's social, economical, agricultural, cultural and political development of the country. The RUSA (Rastriya Uchchatam Sarva Abhyan) is giving some hope in development of the College and its amenities including the Libraries and its activities.

## 13. Conclusion

It is clearly understood that the environment in which modern libraries operate is changing. It is both facing challenges and opportunities. One way of doing that is engaging in Knowledge Management activities, that is creating, capturing, sharing and utilizing knowledge to achieve the library goals. KM is viable means in which libraries could improve their services and become more responsive to the needs of users. People gain more knowledge from their experiences and their peers' expertise. Libraries need to recognize the knowledge, its staff has to create knowledge environment in which their knowledge can be valued and shared. So the Library Professionals should walk along with the domain of Knowledge management like Dr.A.K. Jain Saab.

## References

Award, Elias M.and Ghaziri, Hassan M.Knowledge Management –Delhi: Pearson Education, 2004.pp.1-46. www.ucn.edu/-sunyliu/inls/introduction_to_knowledge_management_html

Malhotra, Y.Knowledge management and new Organization Forms: A Framework for Business Model Innovation. *Information Resources Management journal*, 13(1), 5-14January-March, 2000.

Oosterlink, A and Leuven, K.U.(2002)*Knowledge management in post-secondary Education: Universities*. http://www.oecd.org/pdf/M0027000/M00027356.pdf.

# 18

# Reading Culture and Changing Technological Environment

**Dr. Murali Taddi[1] and Dr. P. Bhaskara Rao[2]**

[1] Course Coordinator and Asst. Professor. [2] Professor (Retd.)
Department of Library and Information Science
Dr. B.R. Ambedkar University, Srikakulam, Andra Pradesh

**ABSTRACT**

*The concept of reading culture was challenged by the impact of ICT or in the changing technological environment; influenced through three important ways viz. 1. Instead of reading a traditional book; most of the users accustomed with: a) Viewing & Listening Habits; b) Internet Use Habits; and c) Media Use Habits. The T.V. and Radio, Motion pictures, Sound Recordings, Video and Audio Recordings, Cassettes and CD-ROMs etc. influenced by the Viewing and Listening Habits. The Internet Use Habits influenced by; Search Engines, Web Browsers, Subject/Content Searching, through various search strategies; Networks and Chatting etc. And the Media Use Habits, influenced by the: Print, Non-print, Non-Book; Special Materials, Microforms, A.V. Materials, and Multimedia etc. This paper attempts to profile the overview of reading culture and changing technological environment.*

***Keywords:*** *Reading culture, Library science, Technological environment, Internet, Social media*

## 1. Introduction

Library and Society are inter-linked and inter-dependent, as well the Society without libraries has no significance, and libraries without society have no origin. J.H. Shera expressed, that, "The Library is a product of our cultural maturation". Thus, the Library is a "Mind of the Society", it reflects man to be self-governing, truth-loving, unity, and non-violence. He should be well-informed, experienced, and become knowledgeable through reading habits, every individual; transformed into a champion of freedom of thought and expression, imaginative, and influence in molding the life of the community. The reading habits among the individuals serve as a vehicle of social progress,

The study felt, that the libraries, and the available type of sources in the library, and the developments in the class room teaching, through teaching aids, are also contributing, some to bring cultural changes in the reading habits among the user community. In the real world, the socio-economic and the political culture of public administration practices, influenced the use of computer tablets directly by the administrators, and as well by the general public or the users can also using the ICT tablets in all sorts of tax payments , all types of user charges payments etc. are also basically contribute a lot to bring change in the minds of public, and are more dependent on the computer gadgets, thus neglecting the reading habits in this information age.

The class room teaching in all levels, i.e. from masters course school level, the academic administrators demanding for the introduction of e-learning concepts, PPT presentation, computer aided class rooms for the school children, and in the central schools, Satellite Class Room Content delivery operations are the basic evidence of impact of ICT in the form of e-Reading and e-class room technologies positively influences towards the electronic gadgets as their prime reading formats, than the printed book. Thus the basic characteristics of the ICT, i.e. Viewing and Listening; Internet Use Habits; and Media Use Habits influences the users, more towards the use of e-resources than the printed book, it shows that most of the users migrating towards the use of ICT technologies.

## 2. Definition and Review of Literature

Effective reading is the most important avenue of effective learning. Reading is interrelated with the total educational process and hence, educational success requires successful reading. Reading is the identification of the symbols land the association of appropriate meaning which it requires identification and comprehension. Die to the advent of T.V., Radio , Internet and the impact ICT (Fig.1), reading habit has lost importance as both the young and the old, who are more attracted towards the multimedia effect of the sound, music, action, motion, color and contextual sceneries.

Reading is a process of thinking, evaluating, judging, imagining, reasoning and problem solving. It includes learning, reflection, judgment, analysis, syntheses and problem solving. It includes attention association, abstraction, generalization, comprehension, concentration and deduction. **Throndike** concluded that the reading of a paragraph gives the value of analytical understanding of that item.

**Gray,** described he reading process, including recognition, understanding, reaction and integration. **Shaw** noted that reading is the process of seeking or perceiving or observing and assimilation of their interrelationships. **Hildreth** expressed that reading requires inference, weighing the relative importance of ideas and meanings. The explosion of knowledge and revolution in Information Communication Technologies (ICT), the developments in Science and Technology brought tremendous change in printing technology in producing volumes of reading materials in digital lightweight portable reading devices, and the computerized tablets and other e-books, e-documents, influencing the user community towards the digital sources, than the printed book.

In this paper the researchers made an attempt to discuss about the important areas, that influence the reading habits in children particularly, adults in general, towards the e-books and e-documents; the impact and role of ICT in promoting the digital sources; and factors ; influencing reading are extensively discussed in this paper.

## 3. Objectives of the Study

- To evaluate the importance of Reading Habits, and the impact of printed book;
- To examine the influence of Information Communication Technologies on Reading Habits;
- To find out the important characteristics of e-documents, while comparing with the traditional book;
- To identify the hindrances to the promotion of reading habits in general.

## 4. Research Methodology

The present article articulated on the review of related literature, and the latest research studies, which are presented in this article. The present article represented the latest research studies and articles published in different countries, from various national and international journals and the research methods used in this article, like the simple averages, followed by the related empirical studied etc. The historical and theoretical comprehensive comparative studies and research methods are projected in developing this article. In this article, the latest library practices, including the changing, technological environment creating diversity in reading the content, which effect the reading habits are also included in this article. The available popular reviews, from various journals and books, through print, non-print, and the web sources are their research methodologies represented are used in framing this article.

## 5. Importance of Reading Habits

The importance of reading habit is termed and understanding as a practice of intrinsic character of human beings. Reading is a process of thinking, evaluating, judging, imaging, reasoning and problem solving. It explores the individual attention, association, generalization, comprehension, concentration and deduction. The concept of Reading Habits, consisting of a grasping, understanding evaluating and thus find out the meaning. Reading is an activity which is very natural, spontaneous existence in a human body. It is also called as a neuro-muscular activity.

## 6. Printed Book & Reading Habits

Book reading habit is an intelligent and wise window of human generations. The reading experience extend and intensify human knowledge and gain deeper understanding about the world. The reading habit hold the most significant place in teaching, research and understanding as a expressive means of communication in the literary society. The book in printed word form, in its traditional conventional form is still, grabs and stands as a most important item. Thus the imparting of quality education and research progress and achievements are conceived through the good reading habits, among the user community or among the individuals. The reading and learning is a self education, and basic tool for education, better understanding and for self discovery. The good reading habits develop the ability of cautious learning, accuracy of information, enthusiastic attitudes,, morals, belief and judgments. The influence, importance, and place occupied by books in the context of reading habits, wherein it provides and cultivate human values and promote education, teaching, research and training. The background history, situation thus appears to be set as the most widely used tool in the historical times of education is the Printed Book. The importance of the reading habits, and the benefits out of reading, particularly in reading the printed book. The brief analytical presentation shows the importance of the book reading.

## 7. Impact of Book on Reading Habits

Here an attempt is made to discuss the importance of book on Reading Habits, while comparing with the non-book, special materials and microforms. Regarding the impact of a book, on reading habits, **Hawkins (2000),** suggested that the advances in production and delivery of e-resources, still the publishing industry continued to print a book according to the required demands. It shows that people read book actively, focusing one or more number of text books, for a variety of reasons. Secondly, the **Schilit (1999)** expressed, that printed books are a long-lasting cultural icons because they are easy to use, generally portable and resistant to damage. Britain recognized as a nation of book lovers with novels and non-fiction books read in 90% of homes. On an average, adults read books for five hours a week and 15% read books, at least 11 hours. Further the printing technology has been optimized step by step over a period of 500 years, and the growth and development is goes on increasing. And the printed book will be continued to maintain its unique and indispensable position within the available new technologies with different media. The studies by Churchill & Johnson (1997) established that the latest technologies are still no match for the book and the new technology has not yet made any impact on people's reading habits. It is stated further, that Surveys in 1989 and 1995 found that 39% of the people reading a book for pleasure within two weeks, and the same number bought more than 16 books a year increased from 28% in 1989 to 30% in 1995. Lastly, **Moyes (2000)** expressed, that the long statements about the death of the book is sill fictional. The above studies exhibits, that the reading habits about the printed book among the general public have its impact still today. But there are some other stories and studies which indicate there are decreasing trends of reading habits in general and reading internet publications have increasing.

## 8. Increasing trends of Online Reading

There are some studies and surveys, which are held between 2000 and 2008, which indicate that the decreasing trends of book reading and increasing trends of internet and online reading. During 2003 and 2005, the China Research Institute of Publishing Science conducted a survey and found that the reading habits of traditional books has fallen among the Chinese, while the number of reading Internet publications has increased sharply. It is also established that the book reading habits among the Americans is also decreasing, the studies found, that those who did not read a single book in a year doubled from 1975 to 1990 (8% to 16%). In another study by **Koskimaa (2003),** found and identified the decreasing trends of reading books; while migrated towards reading of different types of texts, most notably magazines and Internet pages. The **"People's Daily Online, 2007",** conducted studies on popularity of book and reading habit, it was observed that the decreasing trends of book reading i.e. 60.4%(1999); to 51.7%(2003); and 48.7%(2005); which indicate on an average about 11% decrease in six years. It clearly shows, that the popularity of book reading was decreasing, while the online reading habits is growing rapidly, from 3.7% in 1999 to 18.3% in 2003 to 27.8% in 2005. But **Broddason (2006),** conducted number of studies during his 35 years of his research, at Iceland, exhibited, that there is a constant increase of non-readers from 11%(1968) to 33%(2003). This shows the diminishing trends of readers are also applied to books also , i.e. between 8%(1997) to 3%(2003). Broddason expressed, that there us a decreasing trends of book reading; applied to other print media, and it is due to the introduction of internet.

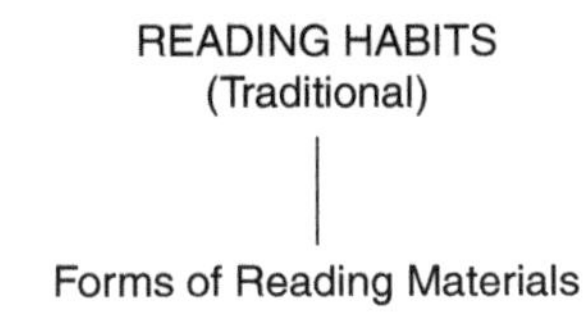


| Printed Book Natural Language | Illustrations (B & W/ Col.); Figures | Charts, Diagrams & Models | Graphic Materials & Different Art forms | Sculpture, Stone & Metal |
|---|---|---|---|---|

Impact of ICT on Reading Habits
(IN THREE BROAD FORMS+)

| 1.Viewing & Listening Habits | 2.Internet Use Habits | 3.Media Use Habits |
|---|---|---|
| T.V<br>FM Radio<br>Video Recording<br>Audio Cassettes<br>CD-ROMs | Search Engines<br>Web Browsers<br>Subject/Content<br>Searching Net works<br>Chatting etc. | Print<br>Non-Print<br>Non-Book Materials<br>Special Materials<br>Microforms<br>A-V Materials<br>Multi Media<br>Technologies |

**Fig.1**

1. Reading Habits (Traditional) most concentrated on books basically, followed by the other items shown in the table in grabbing the user concentration and interests of reading & learning.
2. User's total interests & concentration was split into different forms of e-sources, as the characteristics and forms of the e-materials resembled with the traditional forms, with different modes, and are easy to use. In this process the users concentration and interests of reading was shared between the traditional and non-conventional e-sources.

The impact of ICT on Reading habit develops the mind with new thoughts of intellectual abilities through non- conventional formats (Murali, T & Bhaskara Rao. P, 1916) Sir Richard Steele has logically noted "Reading is to mind and the exercise to body (Cole, 1994). Reading is most acceptable single study tool, and it is next to teacher and text book. Reading is very essential for intentional understanding world brother hood to achieve a the global family

## Impact of New Technologies on Reading Habits

In this study it is felt, that the new Technologies are broadly classified into three types of reading Practices habits, i.e. 1) Viewing and listing habits; 2) Internet Use habits and; 3) Media Use Habits. The technologies involved in these three subject areas are considered as major hindrance to the promoting of Reading habits Fig1 describes the technologies involved in diverting the Reading Habits into "Listing and its effecting on Reading habits and Media Use Habits its effect on Reading habits are as follows:

## 1. Listening and viewing Habits

The Technologies which are part of the ICT, influences the reading Habits, towards the listening and viewing habits. The users or the readers who are gaining knowledge or information are enjoying the experiences of Reading was conceived through Radio, T.V., Motion Pictures Video and Audio recording and printed music by the listening and viewing technologies . The latest studies at Sri Lankan Secondary Schools conducted by Gunasekara (2002) found three major hindrance in promoting the reading habits viz., 1) Preference to chatting and listing to the Radio; 2) Non availability of sufficient reading materials at School Libraries, and, 3) Most of the studies giving preference to viewing the T.V. But in their view, the reading was meant for the examination purpose only; in one investigation held, on primary and middle school children at China reported that "Watching TV was the most popular leisure activity, followed by reading books (Lice, 2000). And a similar survey of primary to secondary students in Hong Kong found that watching T.V and playing computer games were the most preferred School leisure activities (Education Department, 2001). In another study at Britain established, that people aged between 11 and 18 years were more likely to watch T.V or use the internet than reading books (Mori, 2004). There are number of studies exhibits that ICT Technologies in the form of E-resources, e-documents and Educational Computer tablets influences the reading habits among the children in particular are diverted listening & viewing Habits.

## 2. Internet use Habits

The internet and computer Technologies followed by the light weight and portable electronic devices and computer tablets, and dominated in the schools in particular. Now the text book publishers are simultaneously converting the most popular printed titles are produced into E- book formats. The e-books and e- Journals are also well in circulation particularly in Central Schools at 10+2 level in India. "The impact of ICT on reading habits visible through the internet use. The ICT enabled libraries transformed into digital Libraries, where the e- resources are available with multi user facilities, through consortium of e-journals, e-documents and e-resources. These e-resources are available in CD-ROM format available on-line and sometimes, these publications are originally published in print or other formats; concentrated into digital formats to be accessed on Web (Asemi, 2005). The emergence of World Wide Web (WWW) is greatest symbol of giving shift towards the scholarly communication through e-resources (Megara 2002 and 2004). The importance and impact of Web browsers viz., MSN, Net cape, and Internet Explorer and a number of "Search Engines like Yahoo, Google, Hat Bat, Alta Vista" etc. are also important sources of nascent information sources, which will divert the users attention towards the Internet use Habits, than the reading habit.

The e-resources and the digital sources are the basic items of information products, on which most of the user community took a diversion on latest Information and Communication Technologies (ICT). The impact of ICT through Internet Viewing TV., and Search for the subject areas shared the time & space of reading habit, towards the digital formats, accessed through Web (Asemi, 2005). The various forms of ICT, like, Cell phone, TV., Computer and Internet, followed by digital sources e-journals, e-documents and e-books giving direction and causing diversion towards the "Internet use Habits" instead of traditional reading habits. In addition to this the e-resources compose of all types of popular literary manuscript collections, photography's collections of mountains, forests, birds , Animals and Waterfalls etc. may be available in 3D Formats followed by digital formats with multimedia effect have captured, more than half of the time from the reading habits. And it is clearly visible that individuals have reduced their reading contacts from the world of books and other print materials. To assess the impact of ICT on Reading Habits there are numbers of studies have been carried out all over the world, to find out, how far the impact of new technologies or ICT has prevailed over the reading habits, through the Internet use Habits. The review of literature clearly shows, that the "Internet Use patterns/ Habits" in Education, research Library and the information literacy activities in libraries giving positive direction towards the Internet use, over the reading habits.

## 3. Media Use habits

The print media, followed by the non print media non book media microforms A.V. methods, blend with latest technologies of ICT in the name and multimedia about cause reasonable diversion of attention among the users/readers towards the other technologies, other than print media particularly the multimedia among the students indicted attachment and user reading through its non conventional formats, motion, sound and color etc now the libraries in the digital formats of various electronic formats, such as e-books, e-journals and e-documents. Then e-documents are includes in CD-RoM formats as online sources and databases accessed through Web. Most of these sources have had the effect of multimedia. The information literary programme , and IT Training, providing information to the user, how to use the E-resources than the print sources further, the students on media user habits, (Finn, 1980) reported that the average child entering first grade spent 5,000 hours on watching TV and the same child at the age 18 will have spent more time on watching TV than in school. This indicates that the users or students more attracted towards the multimedia effect than the print media. Kamalipur, Robinson & Nortmon ,1998 expressed that students spent 45.05% of their working hours in a week attending to both electronic and print media for their leisure time. On differentiating the use of media and print, that college students exhibited the use of print media (11.09%) more for studying related activities , when the use of media for the leisure time is for outstanding the print media with (27.83%) compared with (3.84%). The study Kaskimara (2003) is of the opinion that reading of books might be decreasing and most of the users reading other types of texts instead, i.e. the magazines and internet pages. The China Research Institution of publishing some surveys, out of which it is found, that number of people in China, reading traditional books has fallen, while the number of people reading internet publications has been increased sharply. This shows that the media use habits are also diverting the user/readers attention towards multimedia not purely on print media.

## 4. Conclusion

The reading habits diverted on ICT products, in this study classified into three broad areas, I.e. 1) Viewing and Reading habits, 2) Internet User Habits, and 3) Media Use Habits. (Fig-1: The conceptual classification of ICT use habits, by Murali,T and Bhaskara Rao,P, 2016). This classification of non-conventional reading habits; creating a huge diversion from and over the traditional reading habits, among the students towards the use of e-books, e- journals and e-sources followed by the internet navigation tools. There are number of studies throughout the world, which it would attribute that the ICT products are more popular in using the non-conventional forms, than the print form. To meet the budget constraints particularly in libraries, the e-resources are introduced in the libraries, followed by the print form. The national government is also more dependent on the digital environments, to improve the quality, transparence, and speed in discharging the work. This e-environment gives the scope of creating big-data for education purpose. Besides that the digital or e-resources are also like a printed book form and these e-resources are also, attributing the similar characteristics of: portability, transferability, longevity & durability, easy accessibility, can get good number of copies, available in reasonable costs. The characteristics of e-resources are very similar to the printed book, the multimedia effect of the e-resources, diverting the user's attention towards the non-conventional sources, than the print forms.

## References

Anderson Karen, (2007) Education and Training for records professional. Records Management Journal,17(2) 94–106

Chea, Y.N. (1998) nurturing the Singapore reader, Reading 32(1)33–34

Cograve, M. S. (2001) Students View on the purpose of reading form three prospective students teacher and parents in the proceedings of the 12th Europe Reading conference Ireland Dublin

Finn, F.L. (1980) TV viewing and reading ERIC Digest Retrieved November 15, 2006 form http:/ / erodigest.org/pre-924 reading.html

Gunasekara, P. Wijeswari. (2002) Education in Srilanka. An overview Colombo Deepani Printers: 22–24

Hangings, C. & Hennery, J.(2006) Reading is a closed book to today's children telegraph. Retrieved 25 may, 2008 form http:/ / www telegraph co uk news/1524595/ reading –is –a book-to –today's –children htmal

Koskimaa, R.(2003) is there a place for digital literature in the information Society? Retrieved June17,2006 form www. Dicthtung –digital org/2003/4-koskimaa.hmt.

Magra, E. (2002) Application of Digital Libraries and Electronic Technologies in Developing Countries: Practical experience for Uganda , Library Review ,51(5): 241–255.

Metsala, JL,&McCann AD (1980) Children's motivations for reading . Reading Teachers, 50(4), 360–361

Moyes, J. (2000) idea that net is killing book reading can be filed under fiction. The Independent may 18 retrieved January 16, 2007 form http://www.independent.co uk/ arts entertainment/books/ news/ idea-that-net-is-killing book reading –can –be-field-under-fictin-718166.html.

Stadler Alosis (1980) children of guest workers in Europe: Social and Cultural needs in relation to Library Services. Library Trends, 29(2): 325–334

Train Briony,(2007) Research on Family Reading: An International perspective. Library Review 56(4): 292–298.

# 19

# Use of RFID Technology in Agricultural Academic Libraries: A Study

**Papegowda M[1] and Dr.Uma Jagannath[2]**

Assistant Librarian, University of Agricultural Sciences
Bangaluru-560 065,Karnataka State
*e-mail; papegowda2009@gmail.com*

**ABSTRACT**

*Radio Frequency Identification technology is the latest technology to make strong security and automatic identification system in the libraries. RFID provides the easier and faster circulation work, security of material, shelf check out and stock verification than Barcode technology. The RFID is the new offspring of the modern technology. The functionality and benefits offered by the RFID systems match the needs and areas of improvement for libraries. In this paper an attempt define type of RFID and advantages and disadvantages.*

***Keyword:*** *RFID, Libraries, RFID Tags, Agricultural Library, Academic Library, Library security*

## 1. Introduction

At present there are 52 Agricultural Universities including five Deemed Universities and two central universities for improving agricultural education. There are 97 research institutions and 589 Krishi Vigyana Kendras involved in full time research and extension activities. All these come under thereifies Indian Council of Agricultural Research (**ICAR**), the autonomous apex body for coordinating, guiding, managing agricultural education, research and extension services in the country. University Library established in 1966-67, is the oldest and biggest university library in the field of Agricultural sciences in Karnataka. Library has splendid collections of Books, Journals. University publications, Government publications, Rare books, Theses/dissertations, Reports, Pamphlets, Maps, Microfilms, Microfiche, CD ROM's/DVD's etc. Provide seamless access to e- resources of world leading online journals databases, such as indiastat.com, Emerald Insight Management Journals , and Free Online Open Access Journals, Biological Abstracts, Food Science and Technology Abstracts , CAB CDs , CABI E-Books, CRC NET E-Books, CERA Consortia, etc, Currently, University library is digitizing its rare collections under e-Granth project aegis of NAIP Consortium. University Library is a partner of Online Computer Library Centre (OCLC) WORLD CAT. With the help of World Cat, users can access the metadata of 200 million documents available in 72,000 Libraries across the globe. The library completely automated with Koha Open Source Software package and integrated with RFID Technology.

In today's information society the librarians have a great responsibility to organize the knowledge centre due to the peak height of information explosion. Libraries are moving towards latest technological environment RFID is a subset of a group of technologies, often referred to as automatic identification, that are used to help machines identify objects and which include bar codes and smart cards. RFID refers to the subset of automatic identification that use radio waves to automatically identify bulk or individual item RFID offer a number of advantages including inventory efficiency, security to library collections and minimal vulnerability to damage.

## 2. Radio Frequency Identification (RFID)

RFID is the reading of physical tag on single products, cases, pallets, or re-usable containers which emit radio signals to be picked up by reader devices. These devices and software must be supported by a sophisticated software architecture that enables the collection and distribution of location-based information in near real time. The complete RFID picture combines the technology of the tags and readers with access to global standardized database, ensuring real time access to up to date information about relevant products at any point in the supply chain. RFID technologies are grouped under the more generic Automatic Identification technologies. Examples of other Auto-ID technologies include smartcards and Barcodes. RFID is often positioned as next generation bar coding because of its obvious advantages over barcodes.

## 3. Historical Development of RFID Technology

RFID was developed in 1948, but its implementations started in 1970s. The first U.S. patent for an active RFID tag with rewritable memory was obtained by Mario E. Cardullo on January 23, 1973. In 1973 Charles Walton, a Californian industrialist, received a patent for a passive transponder that was used to unlock a door without a key. Then Walton licensed the technology to a lock making company called Schlage and RFID technology is another form of automated identification system, which is similar to bar codes. RFID in India was developed in the 1940's for defense applications. 1st time it was used for commercial purpose in 1980 for cattle tracking Applications. Recent interest is in making RFID technology more ubiquitous in the global value chain. The first library suppliers started to market their systems in the mid 1990's. During the 1990's the proliferation of competing systems and radio frequencies employed created the need for standards and interoperability. Libraries need the higher frequency waves to allow for smaller, less powerful and portable readers. As complexities and uses increased, standards were developed to allow systems to work together. Development of standards is still going on with the latest standard being release late in 2004.

## 4. RFID System Components

1. **Tag:** Tag is the heart of the system, which can be fixed inside a book's back cover or directly onto CDs and videos. This tag is equipped with a programmable chip and an antenna each paper-thin tag contains an engraved antenna and a microchip with a capacity of at least 64 bits. There are three types of tags: read only", "WORM," and "Read/write RFID readers or receivers are composed of a radio frequency module, a control unit and an antenna to interrogate electronic tags via radio frequency communication. The reader powers an antenna to generate an RF field.
2. **Antenna:** The antenna produces radio signals to activate the tag and read and write data to it Antennas are the channels between the tag and the reader, which controls the system's data acquisitions and communication. The electromagnetic field produced by an antenna can be constantly present when multiple tags are expected continually. Antennas can be built into a doorframe to receive tag data from person's things passing through the door.
3. **Server:** The server is the heart of some comprehensive RFID systems. It is the communications gateway among the various components (Boss, 2004). It receives the information from one or more of the readers and exchanges information with the circulation database. Its software includes the SIP/SIP2 (Session Initiation Protocol), APIs (Applications Programming Interface) NCIP (National Circulation Interchange Protocol) or SLNP necessary to interface it with the integrated library software but no library vendor has yet fully implemented NCIP approved by NISO The server typically includes a transaction database so that reports can be produced

## 5. Different Types of RFID

Three primary frequency bands are being used for

- Low frequency (125/134 KHz) – Most commonly used for access control, animal tracking and asset tracking.
- High frequency (13.56 MHz) – Used where medium data rate and read ranges up to about 1.5 meters are acceptable. This frequency also has the advantage of not beings susceptible to interference from the presence of water or metals.
- Ultra High Frequency (850 to 950 MHz) – Offer the longest read ranges of up to approximately 3 meters and high reading speeds.

## 6. Use of RFID Technology for Library Services

Use RFID tags on book and other item to provide identification during check-out, check-in, inventory, and for theft deterrence.

- **Check-Out:** Check-out or charging of books is a time consuming process. Users can check-out book on their own with the help of self check-out station. Self checkout station is composed of a proprietary touch-screen device along with an RFID reader plus special software for personal identification, book and other media handling and circulation.[4]
- The LICs to speed up book charging process i.e. checkout several books/ items simultaneously, frees staff for deploying them on other activities, reducing queuing time, and provides user privacy.

## 7. RFID in University of Agricultural Sciences (UAS), Bengaluru

The University Library completely automated with Koha Open Source Software package and integrated with RFID Technology. It is also a time consuming process. Manual discharging can be significantly easier, faster, and more ergonomically friendly with RFID. More books/items can be checked in at one time. The Book Drops system can be useful for this activity. In this system, user inserts the LIC books/items into the slot. The RFID reader captures the electronic signature and sends to backend system for loan cancellation. User's record is updated immediately.

- **Stock Verification**: RFID library system would speed up the finding of books and also improves the stock control of the library. [6] It comprises basically a portable scanner and a base station. High-speed inventory and identify documents which are out of order can be done through the proper use of RFID technology i.e. scan documents on the shelves without removing them.
- **Security at the gate:** When a user leaves the LIC with the issued document these are checked at the exit gate. For this purpose, RFID technology can be efficiently used and a terminal/lane installed on the gate. Theft detection is an integral feature of the chip/circuit within the RFID tag.

It is a stand-alone technology. Each lane is able to track items of about 1 meter. When an unissued document passed through the gate, the terminal/ lane will give an alarm. In this way RFID system can minimize theft of LIC items. In addition to the above application, RFID technology can play an important role in case of retrospective conversion process and inter library loan process.

## 8. Advantages of using RFID in Libraries

- Increased user satisfaction
- Provide reliable statistics for management information system
- Improves-efficiency of the staff and quality of services
- Saves the time of the users;
- Improves information availability

## 9. Disadvantages

- **Lack of standard:** The tags used by library RFID vendors are not compatible even when they conform to the same standards because the current standard only seeks electronic compatibility between tags and readers. The pattern of encoding information and the software that processes the information differs from vendor to vendor, therefore, a change from one vendor's system to the other would require retagging all items or modifying the software.
- **High cost:** The initial investment towards the implementation of RFID technology is very high. The cost of tags range between Rs. 50-60. Besides tags, the reader i.e. exit
- **Censor,** circulation station, scanner, etc. are also quite expensive. These situations discourage many LICs to adopt this technology.
- **Exit sensor problems:** Exit sensor may create problem if there is power failure or if the **book passes** through the side of the pedestal, which is out of the range of the antenna.
- **User privacy concerns:** Privacy concerns associated with item-level tagging is another significant barrier to library barrier to library use of RFID tags. The problem with today's library RFID system is that the tags contain static information that can be relatively easily read by unauthorized tag readers. This allows for privacy issues described as tracking and hot listing. Tracking refers to the ability to track the movements of a book by correlating multiple observations of the book's barcode or RFID tag. Hot listing refers to the process of building a database of books and their associated tag numbers and then using an unauthorized reader to determine who is checking out items in the hotlist. RFID (**radio frequency identification**) is a technology that incorporates the use of electromagnetic or electrostatic coupling in the radio frequency (RF) portion of the electromagnetic spectrum to uniquely identify an object, animal, or person.

## References

Boss. R. W. (2003). RFID technology for libraries [Monograph]. Library Technology Reports. November-December 2003.

Curran, K. and Porter, M. (2007).A primer on radio frequency identification for libraries, Library Hi Tech,25 (4), 595 – 611

Koppel, T. (March 2004). Standards in Libraries: What's Ahead: a guide for Library Professional about the Library Standards of Today and the Future. The Library Corporation. Retrieved from http://www.tlcdelivers.com/tlc/pdf/standardswp.pdf.

Lalit Kumar and Lokender Singh (2013). Use of RFID Technology in Library Security E-Library Science Research Journal, (1) Issue. (3), P 1 – 7.

Lalit Kumar and Lokender Singh Vol.1,Issue.3/Jan. 2013 e-Library Science Research Journal page 1 – 7)

LibBest: Library Management System. Retrieved from http://www.rfid.library.com

RFID in U.S. Libraries. Retrieved from http://www.niso.org/publications/rp/RP-6- 2008.pdf

Sanku Bilas Roy and Moutusi Basak (2011).RFID Technology in Libraries and Information Centers: Beginning of a New Era International Journal of Information Dissemination and Technology, 1 (4) 249 – 252.

# 20

# User's Satisfaction Level of Library Sources and Services and Overview in AKY Polytechnic College, Tirunelveli: A Study

**M.Mani[1] and Dr.A.Thirumagal[2]**

[1]Research Scholar, Manoanmaniam Sundaranar University, Tirunelveli and Librarian
AKY Polytechnic College, Tirunelveli, Tamil Nadu.
e-mail id: mmani.lib@gmail.com
[2]Librarian, Manonmaniam Sundaranar University, Tirunelveli, Tamil Nadu.

**ABSTRACT**

*The study examines the user's satisfaction level of library sources and services and over view in AKY Polytechnic College Library (AKYPCLIB), Tirunelveli. Questionnaire was the instrument for collecting data. 150 copies of questionnaires were distributed, 130 were returned. The present study focuses on users regarding library resources, services, physical facilities, internet, etc.*

***Keywords***: *User Satisfaction, Library Services, Resources, Physical Facilities, AKY Polytechnic College.*

# 1. Introduction

Libraries are constantly increasing their resources and developing new programmes and events to meet the various information requirements of the users. So the library must therefore be alert to change from time to time.[1] Changes demand planning a scientific planning is dependent upon proper evaluation of the present and an accurate forecast of the expected future. Nowadays library is like Knowledge Centre and is the heart of any institute. The user's satisfaction is main motto of library services and it is key success of any library. The services of librarian are also make good customer services and information preferences to ensure that the information needs of users are satisfactorily fulfill within time.[2] A Polytechnic college provides technical education and its important role expected in growth and development of the country so library of polytechnic college library is to satisfy the needs of its users. Libraries assist-in research process by collecting, preserving, and making available an array of information resources relevant to their research community. Present year is an information year and students are not interested in visiting the library physically nowadays and they are more comfortable in retrieving information electronically. The laws of Library Science put forth by Dr.S.R.Ranganathan, Indian Father of Library science, emphasize the importance of books and the significance of readers. The second, third and the fourth laws respectively are 'Every reader his/her book',

'Every book its reader' and ' Save the time of the reader'. Readers are users of the library. Due to information explosion in the modern world, people are craving for knowledge and the libraries around the world are facing a tough challenge to meet their needs[3] but our AKYPC Library takes various steps on improving the library usage and user's statistics. This study will help library authorities to know the satisfaction of users with the existing library services and facilities.

# 2. Profile of the College and Library

AKY Polytechnic College was established in 2015 by AKY Group under the leadership of Mr.A.K.Y Syed Abdul Kadar Chairman-AKYPC. Whose sole ambition is to develop and inculcate technical knowledge and give moralistic education to the students in Tirunelveli and its neighbouring districts. The college strives to impart quality technical education as well as practical skills. It is the objective of the college to inculcate in every student a sense of responsibility towards the society and respect for human life. Besidesdeveloping in them the highest standards of professional behaviour and personal integrity. AKY Polytechnic College is part of the sprawling AKY campus at Tiruneveli 1 km away from Manoanmaniam Sundaranar University. The college was founded with the noble vision to raise professionals and leaders of high academic calibre and unblemished character, nurtured with a strong motivation and commitment to serve humanity. . All the programmes offered by the institute are recognized by the statutory bodies like the All India Council of Technical Education (AICTE), New Delhi and Directorate of Technical Education (DOTE), Chennai.[10]

AKY Polytechnic College (AKYPC) Library supports the institute's program of study and research. Institute archives offer access to a wide range of materials, both printed and electronic. The collection includes many text and reference books, e-books national journals and e-journals, standard newspapers and magazines and online e-resources facility. Library is equipped with all modern facilities at par with international standards. It gives training to our college students to get awareness about the journal collection through open access and also through standard websites such as NPTEL and library network facility like DELNET. The library provides Information Services, Reference Services, Bibliographic Services and Reprographic Service. On a whole, it is a great pride to say that AKYPC library acts as "The Ocean of Knowledge" which quenches the thirst of the library users. The library sends email alerts to the users about various academic activities (FDP, SDP, training programme symposium, workshops, seminars, and conferences). AKYPC Library follows standard quality policy system. Files and registers are maintained well.

## 3. User Study Definition

User Studier is a complicated area of knowledge to define; it can include conventional surveys of library borrowers and this may be the main form of activity which comes to mind when the term 'user studies' is mentioned. The term'user studies' is preferred than 'library surveys' because studies of information needs or information use behaviour focus upon a wider range of information sources and channels rather than simple libraries. 'User Studies' comprises the study of people's need for and use of information. A user study may be defined as a systematic study of information requirements of users in order to facilitate meaningful exchanges between information systems and users.

## 4. Review of Literature

Biradar and Kumar (2000)[4] conducted a study on "Evaluation of Information Services and Facilities offered by DVS Polytechnic College Library: A Case Study". Results shows that 37.5 per cent students and 46.88 per cent teachers are satisfied with lending service and 48.75 per cent students and 50 per cent of teachers respectively have good opinion about book bank facility of the college library.

Busayo (2006)[5] carried a study on "Accessibility and use of Library Resources by Part-time Students: A Case Study of the Federal Polytechnic, Ado-Ekiti, Nigeria". This study investigates the accessibility and use of the Federal Polytechnic Ado-Ekiti library resources by the part-time students of the institution. The result shows that the library is far away from the satellite campus and therefore, not accessible to the majority of the part-time students.

Chinnasamy et al. (2008)[6] made a survey on the usage of electronic resource by management students of Jansons School of Business, Coimbatore. They also studied the impact of electronic resources on the academic work. The study reported that the students used electronic resources for academic and the frequency of use way very high.

Sharma and Singh (2011)[7] Studied the Status of College Libraries in Karnal (Haryana) pertaining to the library collection, membership, library finance, networking, automated system, Internet facilities and other library services are provided in these libraries. The study is based on the survey of college libraries located in Haryana and presents an analytical and comparative study of physical infrastructure, ICT facilities and services.

Balasubramanian and Beula (2012)[8] This study aims to find out the availability of information sources and services available in B.Ed. colleges in Kanyakumari district, Tamil Nadu, India. Majorities of the colleges satisfied the services of the library and suggested to develop the new technologies in the library. The libraries play an important role for supplying information sources to the users. The teacher education college trainees are taken under the study. From this study, it is found that the most the users got timely help form their library staff also demand to develop electronic library. The users are having sufficient knowledge about the print and electronic sources.

Mathurajothi(2012)[9] The study aimed at finding the use of e-resources by the PG students of Gandhigram Rural Institute Deemed University, Gandhigram. 301 questionnaires were distributed among the respondents form seven departments out of which 222 were returned. Results show that 45% of the respondents use internet for educational purposes and 32.9% of the respondents use internet for checking e-mail. Google is the most used search engine and yahoo is their second choice. The purpose for the use of e-resources revealed that most of the respondents are using e-resources for their project work since they have submit the project report as part of their syllabus. 36% of the respondents are using e-resources for preparing seminar presentation.

## 5. Methodology

### 5.1. Research Design

The present study adopts descriptive research design. The purpose of descriptive research is to describe the existing or past state of affairs.

### 5.2. Sample Design

Random Sampling Method is used for selection of sample.

### 5.3. Sources of Data

Both primary & secondary data are used for the study. Secondary data is gathered from books, journals, search engines, etc. Primary data collected from users of the library services.

### 5.4. Tools for Data Collection

Structured questionnaire can be use for collection of data with both open ended and close ended questions.

### 5.5. Sample Size

The sample size for the study is 130.

### 5.6. Analysis of Data:

Simple percentage method is used for analysis of data. Out of 09 tables, 02 important tables are analyzed through Friedman Test by using SPSS 16.0.

## 5.7. Simple Percentage Analysis:

Percentage is used in making comparisons between two or more series of data. Percentages are used to describe relationships.

## 5.8. Friedman Test

The **Friedman test** is a non-parametric statistical test. Similar to the parametric repeated measures ANOVA, it is used to detect differences in treatments across multiple test attempts. The procedure involves ranking each row (or *block*) together, then considering the values of ranks by columns. Applicable to complete block designs, it is thus a special case of the Durbin test.

FM=12bk(k+1)Σi=1k(R¯i−(k+12))2FM=12bk(k+1)Σi=1k(R¯i−(k+12))2

FM∼χ2(k−1)FM∼χ2(k−1).

*FM* statistic follows a Chi-square distribution with κ - 1 degrees of freedom

# 6. Tables and Charts

## 6.1. Table Department Wise Respondents

| S.No | Department | No. of Respondents | Percentage |
|---|---|---|---|
| 1 | Civil | 25 | 19.2 |
| 2 | Mechanical | 50 | 38.5 |
| 3 | Production | 30 | 23 |
| 4 | EEE | 20 | 15.4 |
| 5 | ECE | 5 | 3.8 |
| | **Total** | 130 | 100 |

It is evident from the table and chart 6.1 that 19.2% of the students belong to Civil department, 38.5% belong to Mechanical department, 23% respondents from Production department, 15.4% belong to EEE department and 3.8% from ECE department

## 6.2. Table Frequency of Visit to Library

| S. No | Frequency of visit | No. of Respondents | Percentage |
|---|---|---|---|
| 1 | Every day | 70 | 53.8 |
| 2 | Once in two days | 20 | 15.3 |
| 3 | Once in a week | 20 | 15.3 |
| 4 | Monthly once | 10 | 7.6 |
| 5 | Occasionally | 10 | 7.6 |
| | **Total** | **130** | **100** |

From the table more than 70 respondents (53.8%) visit the library every day, 20(15.3%) respondents visit library once in two days and once in a week and 10 (7.6%) respondents visit the library monthly once and occasionally.

## 6.3. Purpose of Visit to Library

| S. No | Purpose | No. of Respondents | Percentage |
|---|---|---|---|
| 1 | Borrow Books | 50 | 38.4 |
| 2 | Refer Books and Periodicals | 30 | 23 |
| 3 | Browse Internet Resources | 10 | 7.6 |
| 4 | Newspapers Reading | 30 | 23 |
| 5 | Entertainment | 10 | 7.6 |
| | **Total** | 130 | 100 |

It is observed from the table 50(38.4%) of the respondents come to borrow books. To refer books and periodicals 30 (23%) respondents, to borrow internet resources 10 (7.6%), for newspaper reading 10 (7.6%) and the remaining 10 (7. 6%) of them for entertainment purpose.

## 6.4. Satisfaction with Present Collection of the Library

| S.No | Replay | No. of Respondents | Percentage |
|---|---|---|---|
| 1 | Satisfied | 70 | 53.8 |
| 2 | Not Satisfied | 60 | 46.2 |
| | **Total** | **130** | **100** |

It is noticed from the table, 70 (53.8%) of the respondents are satisfied with present collection of the library and the remaining 60 (46.2%) of the respondents are not satisfied.

## 6.5. Uses of Information Sources

| S.No | Documentary & Non-Documentary Sources | No. of Respondents | Percentage |
|---|---|---|---|
| 1 | Books /E-books | 45 | 34.6 |
| 2 | Periodicals | 35 | 26.9 |
| 3 | Dictionary | 20 | 15.4 |
| 4 | Year Books | 10 | 7.6 |
| 5 | Question Bank | 5 | 3.8 |
| 6 | CD/DVD | 15 | 11.5 |
| | Total | 130 | 100 |

From the table, it is observed that most of the 45(34.6%) respondents use books/e-books as main information source, 35(26.9%) respondents use periodicals as information source, 20(15.4%) respondents use dictionary, 10(7.6%) respondents use yearbooks, 5(3.8%) respondents use question bank and 15(11.5%) respondents use CD/DVD as information source

## 6.6. Barriers in Using Library Resources

| S. No | Barriers | No. of Respondents | Percentage |
|---|---|---|---|
| 1 | Library Timings | 50 | 38.5 |
| 2 | Lack of Friendly staff | 10 | 7.7 |
| 3 | Inadequate Resources | 50 | 38.5 |
| 4 | Slow internet connection | 20 | 15.3 |
| Total | | 130 | 100 |

From the table, it is observed that most of the 50 (38.5%) respondents feel that library timings as main barrier, 50(38.5%) respondents feel that resources are inadequate, 20(15.3%) respondents feel that internet connection is slow, 10(7.7%) respond that there is a lack of staff support.

## 6.7. Physical Facilities of the Library

Hypothesis:

$H_0$: Students perception towards physical facility of the library is identical.

### 6.7.1 Descriptive Statistics

| Physical Facilities | N | Mean | SD |
|---|---|---|---|
| Maintenance of Library | 130 | 3.4231 | .93066 |
| Ventilation in the Library | 130 | 3.2000 | .80116 |
| Furniture | 130 | 3.2308 | .89389 |
| Lighting Facility | 130 | 3.0769 | .92019 |
| Reading Room Facility | 130 | 3.3231 | .83727 |
| Drinking Water Facility | 130 | 2.3846 | 1.08109 |

$H_a$: Students perception towards physical facility of the library is different.

### 6.7.2 Friedman Test

Ranks

| Physical Facilities | Mean Rank | Rank |
|---|---|---|
| Maintenance of Library | 4.28 | 1 |
| Ventilation in the Library | 3.73 | 4 |
| Furniture | 3.80 | 3 |
| Lighting Facility | 3.41 | 5 |
| Reading Room Facility | 4.03 | 2 |
| Drinking Water Facility | 1.75 | 6 |

| Test Statistics[a] | |
|---|---|
| N | 130 |
| Chi-Square | 345.991 |
| Df | 5 |
| Asymp. Sig. | .000 |

Since the P value is less than .05 so we reject the null hypothesis ($H_0$). And we concluded that there is significance difference between the students perception towards physical facilities of the library.

Friedman Test indicates the ranks of physical facilities of library. From the table Maintenance of library secures first rank with the mean rank 4.28. Which shows that the students are more satisfied with maintenance of library. Reading room facility places 2 ranks with mean rank of 4.08.Furniture facility places 3rd rank with mean rank of 3.80. Ventilation in the library 4th rank, lighting facility 5th rank, and drinking water facility places last rank with mean rank of 1.75.

## 6.8. Satisfaction Level of Library Services

Hypothesis:

$H_0$: All the services which provided by library is same.

$H_a$: All the services which provided by library is different

### 6.8.1 Descriptive Statistics

| Services offered by the Library | N | Mean | Std. Deviation |
|---|---|---|---|
| Reference service | 130 | 2.8615 | .77514 |
| CAS/SDI | 130 | 3.1154 | 1.01641 |
| Reprographic service | 130 | 3.1538 | .86680 |
| Internet service | 130 | 2.5385 | 1.08659 |
| CD-Rom search | 130 | 3.0538 | .81945 |
| Bulletin Board Service | 130 | 3.2154 | .80680 |
| Circulation services | 130 | 3.3231 | .73890 |
| Newspaper clipping services | 130 | 3.3000 | .73295 |
| Inter Library Loan | 130 | 2.6923 | .82449 |
| Library Orientation Programme | 130 | 3.4154 | .58116 |

### 6.8.2 Friedman Test /Ranks

| Services offered by the Library | Mean | Rank |
|---|---|---|
| Reference service | 4.49 | 8 |
| CAS/SDI | 5.78 | 6 |
| Reprographic service | 5.92 | 5 |
| Internet service | 3.15 | 10 |
| CD-Rom search | 5.44 | 7 |
| Bulletin Board Service | 6.22 | 4 |

| Services offered by the Library | Mean | Rank |
|---|---|---|
| Circulation services | 6.72 | 2 |
| Newspaper clipping services | 6.58 | 3 |
| Inter Library Loan | 3.65 | 9 |
| Library Orientation Programme | 7.07 | 1 |

**Test Statistics**[a]

| | |
|---|---|
| N | 130 |
| Chi-Square | 489.368 |
| Df | 9 |
| Asymp. Sig. | .000 |

Since the P value is less than .05 so we reject the null hypothesis and accept the alternative hypothesis. And we concluded that the satisfaction level of respondents is differ in various service provided by library is differ. It is concluded that most users are satisfied with the Library orientation programme secured first rank, Circulation service scored second rank, Newspaper clipping service scored third rank, Bulletin board service scored fourth rank, Reprographic service secured fifth rank, CAS/SDI service scored sixth rank, CD-ROM Search scored seventh rank, Reference service secured eighth rank, Inter library loan scored ninth rank and Internet service secured tenth rank.

## 6.9. Satisfaction with Overall Facilities and Services

| S.No | Replay | No. of Respondents | Percentage |
|---|---|---|---|
| 1 | Satisfied | 96 | 73.8 |
| 2 | Neither Satisfied nor dissatisfied | 20 | 15.4 |
| 3 | Dissatisfied | 14 | 10.8 |
| Total | | 130 | 100 |

It is noticed form the table that 96 (73.8%) respondents were satisfied, 20 (15.4%) of the respondents were neither satisfied nor dissatisfied and 14 (10.8%) of the respondents were dissatisfied with regard to overall facilities and services

## 6.10. Overview of AKYPC Library Activity

We have conducted library hour in our college to make our students to use library, digital library, e-resources and internet. It is very useful for our students and creates awareness about general knowledge, aptitude and reasoning etc., to develop motivation among the students.

### 6.10.1 Library Hour Classes

We create sapling plantation awareness among the students' community and planted saplings around the college campus. Our Library conducted Inter campus programme like FDP, SDP, Librarian Day, Library Week Celebration, World Book Day, and Training Programme and book fair.

### 6.10.2 Sapling Plantation awareness

In 2015, we conducted the following programmes; Faculty Development programme on "Effective Skill Techniques of Teaching and Learning Process". It is very essential for our staff to develop skill techniques Teaching and Learning Process. Due to which all our staff start to present papers and published papers.

### 6.10.3 FDP Programme

Also we conducted Student Development Programme on Skill Development of LSRW (Listening, Speaking, Reading, and Writing). It is very important for our student to develop LSRW skills.

Our library also conducts two training programmes each semester to create e-mail account and how to usage DELNET and NPTEL and E-literacy Class and recent technologies. We celebrated Library Week Celebration in 2015 at four schools and our college in a grand manner. This programme helped students to improve reading habit and awareness about library and library usage. In this celebration, we had conducted competition on speech, essay, drawing, quiz, etc for students. The participant's students were given certificates and books. We also helped their library management and helped students to become a member in public libraries

### 6.10.4 Library Week Celebrations

We also gave awareness to students to present papers and participate in both state and national level workshops and conferences and also guide them to publish papers and mini projects and made them to attend implant training and department membership in its association. We have created a separate websites for our college library also created a separate Blog and FB Page for our college library to display our college news and library events. I hope that In future our management will conduct workshops, seminars, conferences, and symposium. These kinds of activities are conducted by AKYPC library.

## 6.11. Major Findings of the Study

- Most of the respondents belong to Mechanical and Production department.
- Around 53.8% respondents visit the library every day.
- The main purpose of visiting the library is borrowing books.
- 53.8% respondents are satisfied with the present collection in the library.
- Most of the respondents collect the information form books/e-books.
- The main barriers are library timing and inadequate resources.
- From Friedman test analysis of physical facilities, maintenance of library secured the first rank.
- Most of the respondents are satisfied with the library orientation programme.
- 73.8% respondents are satisfied with the overall facilities and services in the library.

## 6.12. Suggestions and Conclusion

The main purpose of any library is to provide relevant and up-to-date materials with a view to satisfy the information needs of users. It is encouraging to know that the student communities are awakened a lot. The role of librarian in a Polytechnic College is really challenging. Based on the findings it is clear that the library users are satisfied with most of the facilities in the AKY Polytechnic College Library, Tirunelveli. On the whole, the study revealed that information resources, physical facilities and services influence user's satisfaction. Certain remedies can be done such as; to improve the present collection, increase in the speed of internet connectivity, and also increase the Inter Library Loan services.

## References

Balasubramanian, B. and Catherin Beul, C. (2012). Information Sources and Services in B.Ed. Colleges in Kanyakumari District: An analytical Study, *IJISS*, 2(1): 43-44.

Biradar,B.S and Kumar,.B.(2000). Evaluation of Information services and facilities offered by DVS Polytechnic College Library: A case study, *Library Herald*, 38(2): 112-121.

Busayo,I.O.(2006). Accessibility and use of library resources by part-time students: A case study of the federal polytechnic Ado-Ekiti, Nigeria, *Library Review*, 55(2):148-156.

Chinnasamy, K. et al. (2008).E-resources Usage by Management Institute Students: A Study, *Pearl: Journal of Library and Information Science*, 2(I):37-43.

Mani.M and Thirumgal,A. (2015).User's Satisfaction Level of Library Sources and Services in Thamirabharani Engineering College, Tirunelveli: A Study, *PRAGNAVANI Journal*, 10(1):43-53.

Mathurajothi.S. (2012). Access of e-resources by PG students of Gandhigram Rural Institute -Deemed University, Gandhigram, Tamil Nadu, India, *IJISS*, 2(1):45-49.

Sadu Ranganadham and Surendra Babu,K.(2013).User Satisfaction on Library Resources and Services in B.M.S College of Engineering, Bangalore: A Survey. *International Journal of Library and Information Studies*, 3(2):39-47.

Saini,P K., Rajkumar Bhakar and Bhoop singh.(2014). User Satisfaction of the Students of Engineering College: a Case Study of Engineering College Libraries of Jaipur, Rajasthan, *International Journal of Emerging Research in Management & Technology (IJERMT)*, 3(9):16-26.

Selvaraj, A. D., and G. Rathinasabapathy. "A study on electronic information use pattern of faculty members of self-financing engineering colleges in Tiruvallur District, Tamil Nadu." Asian Journal of Library and Information 6 (2015): 3-4.

Selvaraj, A. D., and G. Rathinasabapathy. "Information Use Pattern by Students of Self-financing Engineering Colleges in Tiruvallur District, Tamil Nadu: A Study"." Journal of Library, Information and Communication Technology 6.3-4 (2015): 45-54.

Sharma and Singh. (2011). Status of Information Communication Technology Infrastructure for computerization of College Library Services in the state of Assam, Evaluative Study of College Libraries of Barak Valley, South Assam, *Library Progress (International)*, 32(1):29-59.

# 21

# Implementation of Koha Integrated Library Management System at FC&RI, Ponneri: A Case Study

**M.S.Premraj**

Fisheries College and Research Institute
Tamil Nadu Fisheries University
Ponneri, Tamil Nadu

**ABSTRACT**

*This paper appraised the implementation process of Koha Integrated Library Management System at Fisheries College and Research Institute, a constituent college of the Tamil Nadu Fisheries University, Ponneri. The software installation, data entry and data migration were successfully done and usage of the software began instantly and this paper attempts to document the process of implementation of KOHA ILMS.*

***Keywords:*** *Koha, ILMS, Library Automation, Open Source Software, OPAC, FCRI, TNFU*

## 1. Introduction

Library is a growing organism. Automation is a technique to make a system automated, i.e. self service. A properly computerized library will help its user with quick and prompt predominantly by computerization. Thus library automation means the application of machines to perform different routines, such as repetitive and clerical job involved in the function and services of the libraries. Library automation is concerned with managing controlling and automating library collection, activities and services in an automated library, this software used in most of the activities.

## 2. Fisheries College and Research Institute – A case study

The term Information Technology is made of two separate terms information and Technology where Information is the power and refers to fact and opinions provided and received during the course of daily life. This paper provides information of application of the Koha integrated library system in FC&RI library as a means to acquire, store, transmit, retrieve and process information for its students and staff. It also describes various challenges faced by library staff in implementing this software. The institute is just three years old, but it has been amazingly successful in attracting a number of students for the course of Fisheries. Presently the Institute is running 3 full time courses. The library is using Koha : Integrated library System for automating its library resources and services. The library holds the details of more than 1500 books and serves 12 students and 19 staff.

## 3. Koha Integrated Library Management System (Koha ILMS)

Koha is a full featured Integrated Library System (ILS). there is no cost for the license, you have the freedom to modify the product to adapt it to your needs, etc. Developed initially in New Zealand by Katipo Communications with Horowhenua Library Trust. It is currently maintained by a dedicated team of software providers and library technology staff from around the globe. That by adopting it, the customer becomes "joint owner " of the product. In particular, the customer can freely install new versions or not, and can take part in new developments by financing them or by carrying them out them self.

## 4. Features of Koha ILMS

Koha is an award winning and free/Open source software (no license fee) and the features of this integrated library management system are:

- OS independent any operating system. Linux, Unix, Mac.
- Web based. Web-based Interfaces. We can integrate with website.
- Full MARC21 and UNIMARC support for professional cataloguing.
- Multilingual and multi-user support
- Library-Standards-Compliant. industrial standards & protocols. Z39.50 server.
- Customizable web based OPAC circulation system.
- Online reservation.

- Full catalogue, circulation, acquisitions, library stock management.
- Web based OPAC, public to search the catalogue.
- Major industry-standard database type (text, RDBMS), SQL,MYSQL.
- Serial management module.

Considering the significance of library automation and the easy to use and customization features of Koha ILMS, it has been decided by the authorities to automate the library using Koha ILMS. The Koha ILMS was implemented in the Library of FC&RI, Ponneri, a constituent college of the Tamil Nadu Fisheries University, Ponneri. The Koha Home page and Administration Screen shot is depicted in Figure-1.

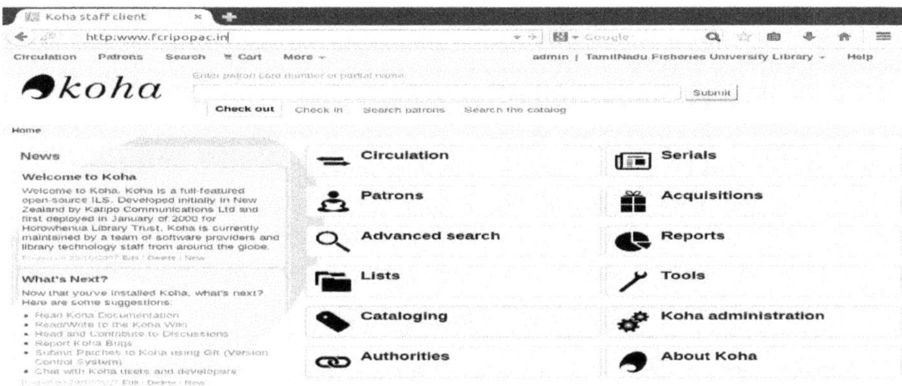


**Fig.1 KOHA – Home Page / Administration Screen Shot**

Home page of FC&RI, Ponneri shows that all the modules are available this front page circulation, patrons information, search for all catalogues, Reports, koha administrations link, tools.

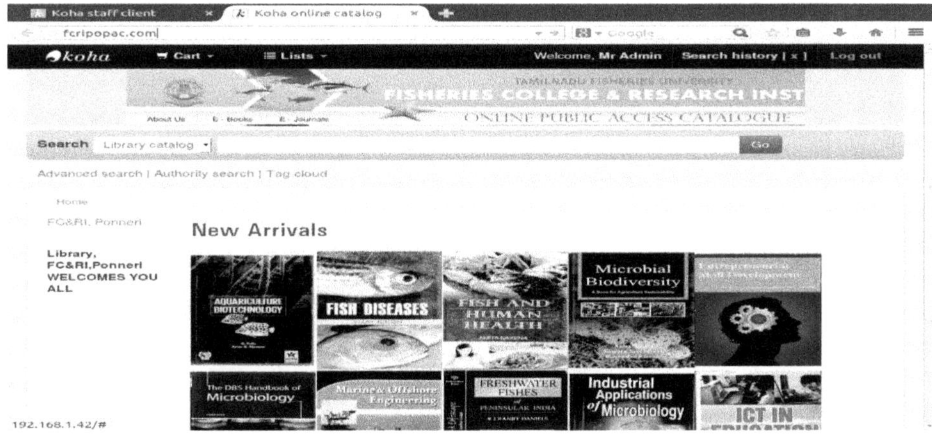


**Fig. 2. KOHA – OPAC Page Screen Shot**

Koha OPAC describes all the book details available in FC&RI library, and also we can access our OPAC page at http://www.fcripopac.com

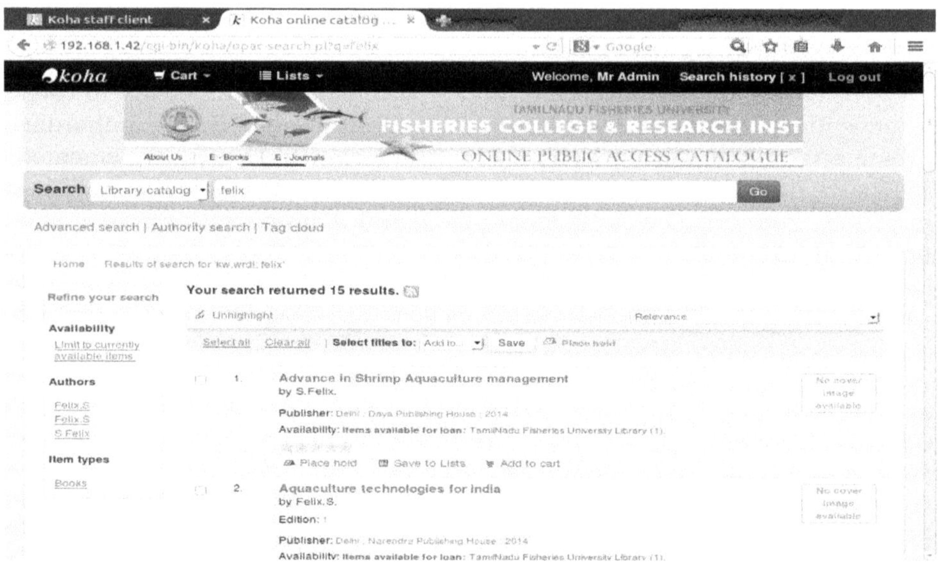


**Fig. 3 KOHA – Author Search Page**

We can view the book title and the author of the book by the boolean search as shown in the above screen shot.

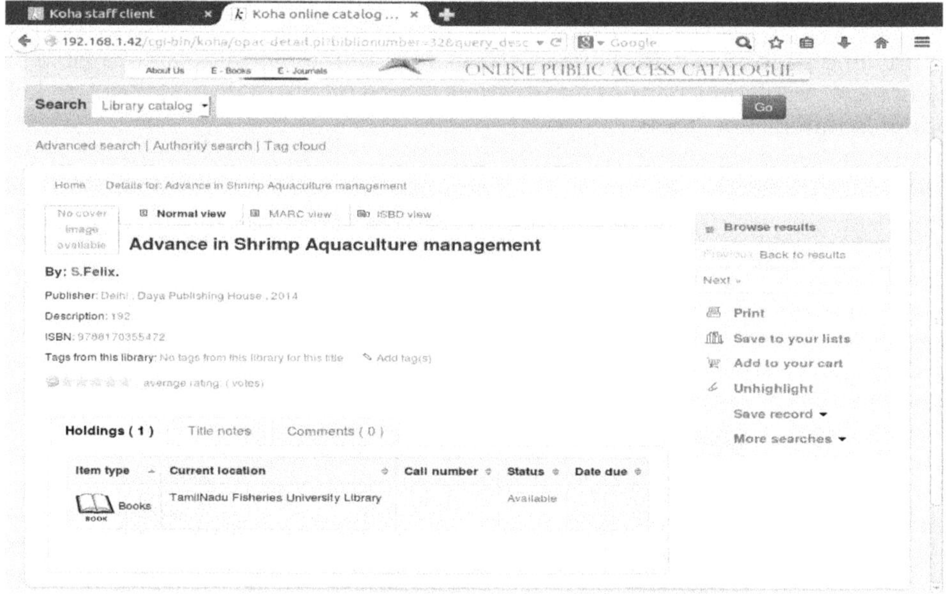


**Fig.4 KOHA – Full Description of the book – Screen Shot**

We can view the full description of the book as shown in the above diagram.

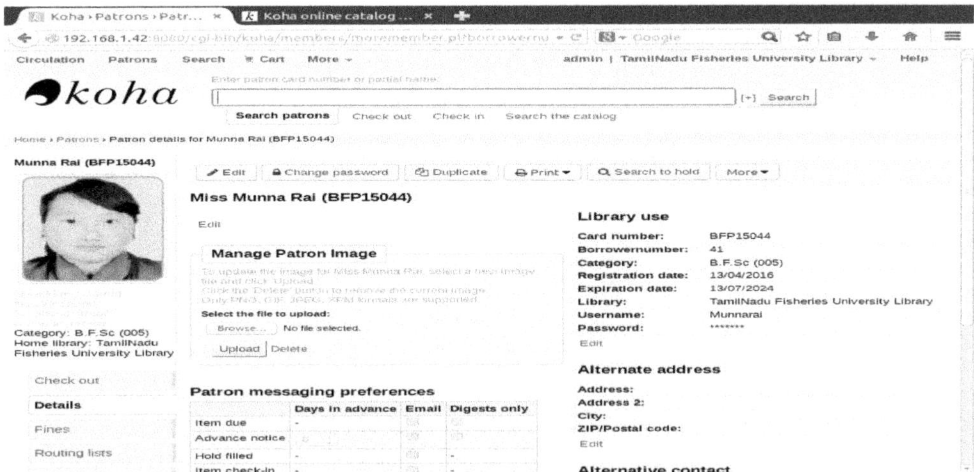


**Fig.5 KOHA- Student and Staff – Circulation Module**

The above screen shot shows the patron details and also the checkout details.

## 5. Conclusion

Koha has been attracting interest from many libraries throughout the world but in India also it is being implemented in many libraries. Koha open source software system has been chosen for the automation of major day to day activities of the Fisheries College and Research Institute. This implementation project had the basic objective of designing a bibliographic database for the Fisheries College and Research Institute with which the automation of circulation routines is carried out. From this point of view it may be concluded that Koha is a useful package for the creation of a database and for information retrieval and for the complete automation of library of any size. Approximately all ILs software offers the same module for all type of libraries, ignoring the aspect of library collection, user and services, but open source ILs software comes with core functional modules, such as online catalogue, circulation, cataloguing with choice of limiting parameters format. The students and staff of FC&RI, Ponneri are effectively using the Koha OPAC.

## References

Bissels, Gerhard. "Implementation of an open source library management system: Experiences with Koha 3.0 at the Royal London Homoeopathic Hospital." *Program* 42.3 (2008): 303-314.

Egunjobi, R. A., and R. A. Awoyemi. "Library automation with Koha." *Library Hi Tech News* 29.3 (2012): 12-15.

Kohn, Karen, and Eric McCloy. "Phased Migration to Koha: Our Library's Experience." *Journal of Web Librarianship* 4.4 (2010): 427-434.

Rathinasababapathy, G. "Strengthening of digital library and information management under NARS (e-Granth): An Overview *Journal of Library, Information and Communication Technology* 4.1-2 (2015): 73-78.

Singh, Manisha, and Gareema Sanaman. "Open source integrated library management systems: Comparative analysis of Koha and NewGenLib." *The Electronic Library* 30.6 (2012): 809-832.

# 22

# Research Output of PSG College of Technology, Coimbatore (1978 -2016): A Scientometric Analysis

**Arumugam.J and Prakash.M**

PSG College of Technology, Coimbatore, Tamil Nadu

**Abstract**

*This paper analyses the growth of scholarly publications of PSG College of Technology, Coimbatore during 1978-2016 with reference to the Scopus international multidisciplinary bibliographical database. The study measures the performance based on several parameters, publishing trend, impact factor, authorship pattern, institutional collaboration of authors, affiliated institutions of authors, and keyword analysis. The number of articles being published in online Scopus database during the period 1978-2016 is 3678. There is an upward trend in number of collaborated papers.*

***Keywords:*** *PSG College of Technology, Scholarly Publications, Scientometric analysis, Authorship pattern*

## 1. Introduction

Scientometrics is a branch of science. These techniques are being used for several purposes in science like evaluation of scientific output by analysing author collaboration, output of publication, authorship pattern, etc., can be measure by this technique. Scientometricians explain about input and outputs of resource in terms of organizational structure. They develop benchmarks to evaluate the quality of information resources and packages of information for decision making in science. It provides a key opportunity to the researcher to publish their articles with new strategies, innovations, new methods and new ideas .In the present study; we did the Scientometric analysis of Electrical Engineering Research, a significantly growing area of science discipline.

## 2. Review of Literature

Subbiah Arunachalam (2001), evaluated the research output of mathematics in India, during the period 1988-1998. Result of this study shows that Indian researchers of mathematical research contribution increased during the study period.

Dutt, Garg & Bali (2003) analyzed 1317 papers published in the first fifty volumes of the International journal of Scientometrics during 1978 to 2001. They found that the U.S.A share of papers is constantly declining while that of the Netherlands, India, France and Japan is on the rise. The research output is highly scattered as indicated by the average number of papers per institution.

Wen et al. (2007) in a survey entitled "Scientific production of electronic health record research, 1991–2005" came to the conclusion that numbers of published articles have significantly increased compared to each 5-year period. Most articles were published in English (98%) and were from the region of America (57%). The top 10 of the 374 journals accounted for 41% of the number of published articles. An analysis of the number of articles related to population showed a high publication output for relatively small countries like Switzerland, Netherlands, and Norway. Generally, they found a considerable increase in the literature of "electronic health research" during 1991to 2005.

## 3. Scope and Methodology

The present study attempts to find out the publication pattern of scholarly articles by the faculty of PSG College of Technology, Coimbatore. The study is based on the references and aims to analyse quantitatively the growth and development of scholarly publications in terms of publication output as reflected in Scopus database during the period 1978-2016.

## 4. Objectives of the study

The following objectives were formulated for the present study:

- To sketch the year wise distribution
- To examine the authorship pattern of the contribution.
- To find out the degree of collaboration
- To measure the source wise publications
- To find out the research productivity in institution wise
- To observe the citation factor of publications.
- To examine the document wise distribution.

## 5. Methodology

This information about the publications has been downloaded from the online SCOPUS database. In all 3678 records were obtained for the period of (1978-2016). Microsoft Excel 2010 was used for analysing data.

### 5.1. Co-authorship and Degree of Collaboration

Co-authorship is a kind of scientific collaboration in which two or more authors share their ideas, resources and data to create a joint work (Inzelt et al, 2009; Nikzad, 2012). Thus, it is essential to measure the intensity of co-authorship trends through the degree of collaboration among the authors of PSG College of Technology. The degree of collaboration among researchers of PSG College of Technology can be calculated by using Subramanyam's formula (1983) as:

DC=Nm/Nm+Ns

Where ,

DC= degree of collaboration.

Nm =number of multiple authored papers

Ns=number of single authored papers

Here, DC= 3250/ (3250+144) =0.95 Since DC value is greater than 0.5 and tends to 1, it is evident that multi authored articles occupy the prominent position indicating the supremacy of team research carried out by the faculties of PSG College of Technology.

## 6. Analysis and Discussion

### 6.1. rowth of Electrical engineering research output:

Table 1 indicates the growth of scholarly articles of PSG College of Technology during 1978-2016. The highest output during the 5 year period was in the year 2014 (11.29%) and the lowest (0.08%) in 1984 & 1987. Further the year wise distribution during the years 2015(11.02%), 2013 (9.26%), 2012 (10.39%) indicates that these years were relatively more productive in relation to total number of scholarly articles of PSG College of Technology.

**Table 1- Year wise Distribution**

| Year-wise | No. of Articles Published | Percentage of 3678(%) |
|---|---|---|
| 2015 | 401 | 11.02 |
| 2014 | 411 | 11.29 |
| 2013 | 337 | 9.26 |
| 2012 | 378 | 10.39 |
| 2011 | 295 | 8.11 |
| 2010 | 226 | 6.21 |

| Year-wise | No. of Articles Published | Percentage of 3678(%) |
|---|---|---|
| 2009 | 235 | 6.46 |
| 2008 | 190 | 5.22 |
| 2007 | 211 | 5.80 |
| 2006 | 171 | 4.70 |
| 2005 | 132 | 3.63 |
| 2004 | 90 | 2.47 |
| 2003 | 48 | 1.32 |
| 2002 | 38 | 1.04 |
| 2001 | 28 | 0.77 |
| 2000 | 29 | 0.80 |
| 1999 | 19 | 0.52 |
| 1998 | 19 | 0.52 |
| 1997 | 15 | 0.41 |
| 1996 | 14 | 0.38 |
| 1995 | 10 | 0.27 |
| 1994 | 8 | 0.22 |
| 1993 | 12 | 0.33 |
| 1992 | 9 | 0.25 |
| 1991 | 7 | 0.19 |
| 1990 | 4 | 0.11 |
| 1989 | 5 | 0.14 |
| 1988 | 4 | 0.11 |
| 1987 | 3 | 0.08 |
| 1986 | 4 | 0.11 |
| 1985 | 7 | 0.19 |
| 1984 | 3 | 0.08 |
| 1983 | 5 | 0.14 |
| 1982 | 3 | 0.08 |
| 1981 | 7 | 0.19 |
| 1980 | 4 | 0.11 |
| 1979 | 6 | 0.16 |
| 1978 | 6 | 0.16 |
| Total | 3678 | 100.00 |

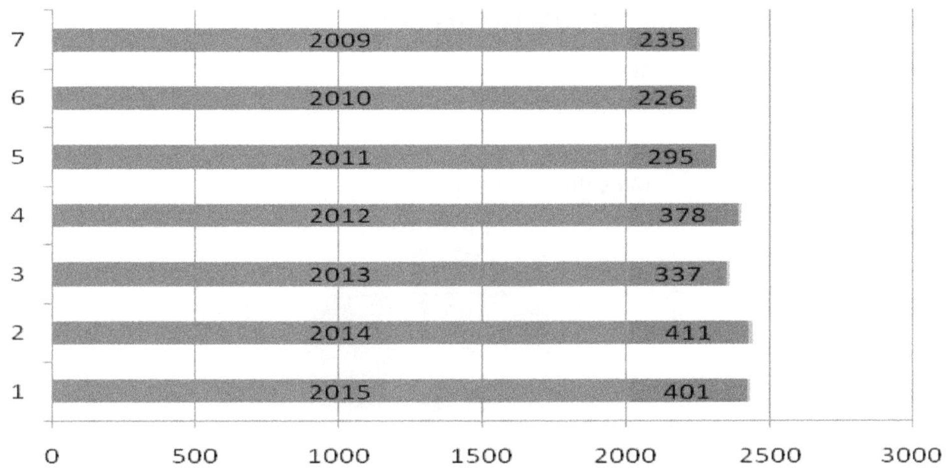


**Figure1- Year wise Distribution**

## 6.2. Authorship Pattern

To find the authorship pattern, entire data was divided in to seven blocks as single, two, three, four, five, six and more than six authored publications as shown in Table 2. The results presented in Table -2 shows that two authored publications were maximum at 1405 (38.61%) more than six authored papers at 106(2.91%) were the lowest.

**Table 2- Author wise Distribution**

| Author wise Pattern | No of records | Percentage of 3678(%) |
|---|---|---|
| Single Author | 148 | 4.07 |
| Co Authors | 1405 | 38.61 |
| Three Authors | 994 | 27.32 |
| Four Authors | 545 | 14.98 |
| Five Authors | 298 | 8.19 |
| Six Authors | 143 | 3.93 |
| More than Six Authors | 106 | 2.91 |
| | 3678 | 100.00 |

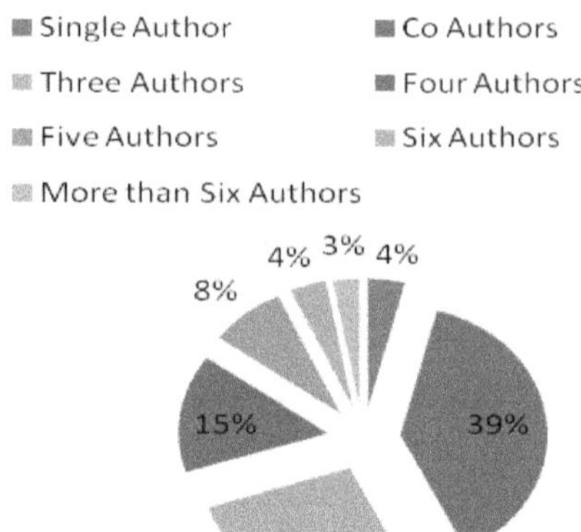


**Figure 2- Author wise Distribution**

## 6.3. Degree of Collaboration:

The Degree of collaboration of authors by year wise is shown in Table 3. The extent of Degree of Collaboration by year wise has been calculated. Accordingly, the degree of collaboration has been calculated for the years 1978-2016. The year wise Degree of collaboration falls between 0.67 and 1.00

**Table 3- Degree of Collaboration**

| Year | Single Author | Two Authors | Three Authors | More than Three Authors | NM | Degree of Collaboration | Total |
|---|---|---|---|---|---|---|---|
| 1978 | 0 | 3 | 3 | 0 | 6 | 1.00 | 6 |
| 1979 | 1 | 5 | 0 | 0 | 5 | 0.83 | 6 |
| 1980 | 0 | 1 | 0 | 3 | 4 | 1.00 | 4 |
| 1981 | 0 | 3 | 0 | 4 | 7 | 1.00 | 7 |
| 1982 | 1 | 0 | 0 | 2 | 2 | 0.67 | 3 |
| 1983 | 0 | 1 | 0 | 4 | 5 | 1.00 | 5 |
| 1984 | 1 | 2 | 0 | 0 | 2 | 0.67 | 3 |
| 1985 | 1 | 3 | 2 | 1 | 6 | 0.86 | 7 |
| 1986 | 0 | 3 | 1 | 0 | 4 | 1.00 | 4 |
| 1987 | 0 | 3 | 0 | 0 | 3 | 1.00 | 3 |
| 1988 | 0 | 3 | 1 | 0 | 4 | 1.00 | 4 |
| 1989 | 0 | 3 | 1 | 1 | 5 | 1.00 | 5 |
| 1990 | 0 | 1 | 3 | 0 | 4 | 1.00 | 4 |
| 1991 | 1 | 4 | 2 | 0 | 6 | 0.86 | 7 |
| 1992 | 0 | 4 | 1 | 4 | 9 | 1.00 | 9 |

| Year | Single Author | Two Authors | Three Authors | More than Three Authors | NM | Degree of Collaboration | Total |
|---|---|---|---|---|---|---|---|
| 1993 | 3 | 0 | 2 | 7 | 9 | 0.75 | 12 |
| 1994 | 1 | 5 | 2 | 0 | 7 | 0.88 | 8 |
| 1995 | 0 | 2 | 3 | 5 | 10 | 1.00 | 10 |
| 1996 | 2 | 4 | 4 | 4 | 12 | 0.86 | 14 |
| 1997 | 2 | 2 | 8 | 3 | 13 | 0.87 | 15 |
| 1998 | 1 | 5 | 8 | 5 | 18 | 0.95 | 19 |
| 1999 | 1 | 10 | 4 | 4 | 18 | 0.95 | 19 |
| 2000 | 2 | 6 | 14 | 7 | 27 | 0.93 | 29 |
| 2001 | 3 | 11 | 6 | 8 | 25 | 0.89 | 28 |
| 2002 | 2 | 13 | 13 | 10 | 36 | 0.95 | 38 |
| 2003 | 4 | 19 | 10 | 15 | 44 | 0.92 | 48 |
| 2004 | 9 | 34 | 25 | 22 | 81 | 0.90 | 90 |
| 2005 | 15 | 48 | 36 | 33 | 117 | 0.89 | 132 |
| 2006 | 11 | 50 | 48 | 62 | 160 | 0.94 | 171 |
| 2007 | 6 | 100 | 53 | 52 | 205 | 0.97 | 211 |
| 2008 | 7 | 68 | 49 | 66 | 183 | 0.96 | 190 |
| 2009 | 6 | 99 | 53 | 77 | 229 | 0.97 | 235 |
| 2010 | 10 | 76 | 68 | 72 | 216 | 0.96 | 226 |
| 2011 | 6 | 130 | 72 | 87 | 289 | 0.98 | 295 |
| 2012 | 11 | 153 | 103 | 111 | 367 | 0.97 | 378 |
| 2013 | 15 | 123 | 93 | 106 | 322 | 0.96 | 337 |
| 2014 | 7 | 156 | 125 | 123 | 404 | 0.98 | 411 |
| 2015 | 15 | 148 | 124 | 114 | 386 | 0.96 | 401 |

## 6.4. Preferred Journals by the Faculty of PSG College of Technology

Table 4 reveals the highly preferred journals by the researchers of PSG College of Technology out of the 3678 articles, 90 papers have been published in "International Journal of Applied Engineering Research", 68 in "Indian Journal of Fiber And Textile Research" and also 66 in "Asian Textile Journal "and so on. Top 10 refereed journals listed here based on the maximum number of publications by the faculty of PSG College of Technology.

**Table 4- Source wise Distribution**

| S.No | Source Title | No. of. Articles Published | Percentage of 3678 |
|---|---|---|---|
| 1 | International Journal of Applied Engineering Research | 90 | 2.45 |
| 2 | Indian Journal of Fiber and Textile Research | 68 | 1.85 |

| S.No | Source Title | No. of. Articles Published | Percentage of 3678 |
|---|---|---|---|
| 3 | Asian Textile Journal | 66 | 1.80 |
| 4 | Indian Journal of Dermatology Venereology and Leprology | 60 | 1.63 |
| 5 | European Journal of Scientific Research | 59 | 1.61 |
| 6 | Studies in Computational Intelligence | 48 | 1.31 |
| 7 | Journal of Scientific and Industrial Research | 38 | 1.03 |
| 8 | Journal of the Institution of Engineers India Part TX Textile Engineering Division | 38 | 1.03 |
| 9 | Man Made Textiles in India | 37 | 1.01 |
| 10 | Applied Mechanics and Materials | 35 | 0.95 |

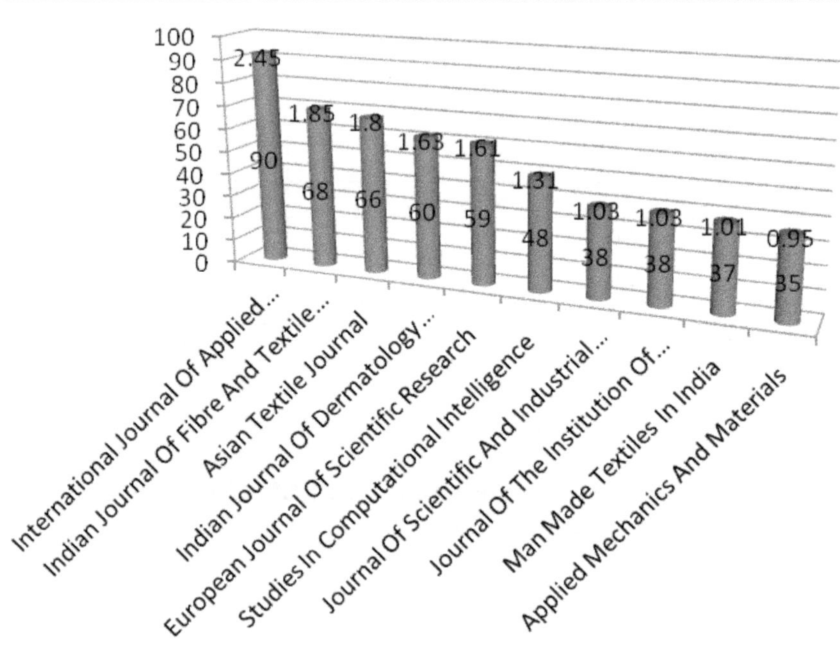


**Figure 3- Source wise Distribution**

## 6.5. Citation analysis

During the period (1978-2016) PSG College of Technology has produced a total of 3678 articles. The highest number of citation was (540) in International Journal of Advanced Manufacturing Technology, (427) Journal of Materials Processing Technology, (390) Indian Journal of Fibre and Textile Research, (381) Indian Journal of Dermatology, Venereology and Leprology and respectively so on. International Journal of Advanced Manufacturing Technology has highest number of citations compared with all other Journals.

**Table 5– Citation analysis**

| S.No | Name of the Journal | No. of citations |
|---|---|---|
| 1 | International Journal of Advanced Manufacturing Technology | 540 |
| 2 | Journal of Materials Processing Technology | 427 |
| 3 | Indian Journal of Fibre and Textile Research | 390 |
| 4 | Indian Journal of Dermatology, Venereology and Leprology | 381 |
| 5 | Journal of Colloid and Interface Science | 356 |
| 6 | Journal of Hazardous Materials | 326 |
| 7 | Materials and Design | 251 |
| 8 | Journal of Industrial Textiles | 240 |
| 9 | Bioresource Technology | 215 |
| 10 | Renewable and Sustainable Energy Review | 200 |
| 11 | Indian Journal of Medical Microbiology | 196 |
| 12 | Dyes and Pigments | 196 |
| 13 | Applied Mathematical Modelling | 174 |
| 14 | Carbohydrate Polymers | 159 |
| 15 | Computers and Structures | 153 |
| 16 | Computers in Biology and Medicine | 147 |
| 17 | Introduction to Fuzzy Logic using MATLAB | 145 |
| 18 | Journal of the Institution of Engineers (India), Part TX: Textile Engineering Division | 145 |
| 19 | Journal of Physics D: Applied Physics | 144 |
| 20 | Materials Science in Semiconductor Processing | 143 |

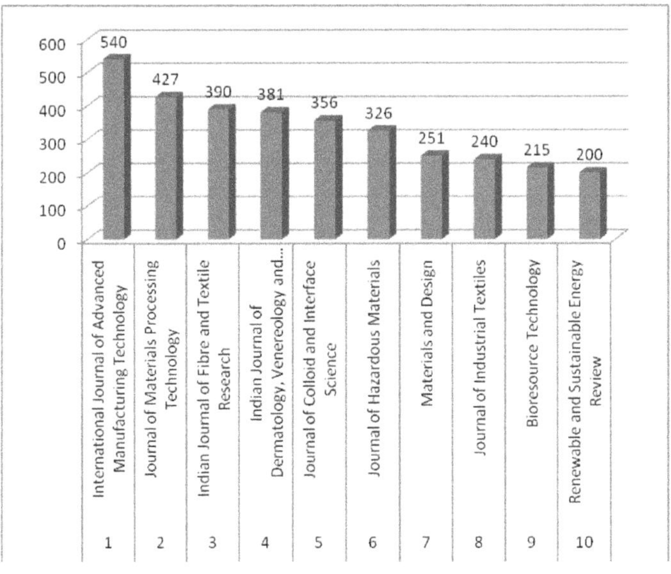


**Figure 4– Citation analysis**

## 6.6. Documentwise Distribution

During the period (1978-2016) PSG College of Technology has produced a total of 3678 articles. The highest number of document type published was articles 2824 followed by conference papers 563, reviews 118, and so on. The least number of articles books 4, short survey 3 and Erratum 2.

**Table 6– Document wise distribution**

| Sl.No | Document type | No of publications |
|---|---|---|
| 1 | Article | 2824 |
| 2 | Conference Paper | 563 |
| 3 | Review | 118 |
| 4 | Letter | 55 |
| 5 | Article in Press | 47 |
| 6 | Book Chapter | 38 |
| 7 | Note | 12 |
| 8 | Editorial | 6 |
| 9 | Book | 4 |
| 10 | Short Survey | 3 |
| 11 | Erratum | 2 |

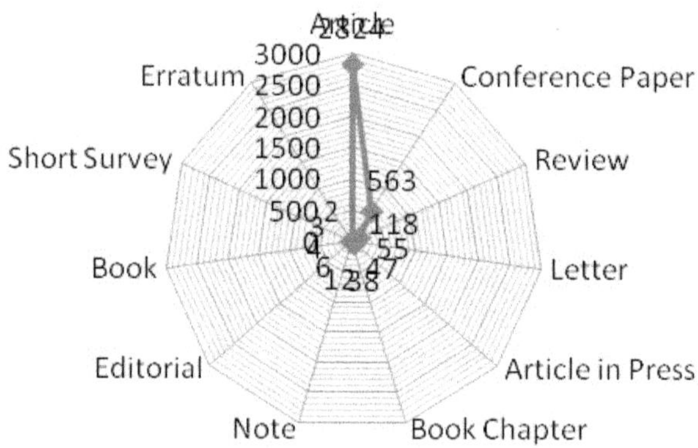


**Figure 5– Document wise distribution**

## 6.7. Collaborative Institutions

Table 7 reveals the highly collaborative institution with PSG College of Technology is Coimbatore Institute of Technology (141), Coimbatore, Anna University, Chennai (97), Kumaraguru College of Technology, Coimbatore (89), and so on. Top 10 collaborative institutions listed here with the faculty of PSG College of Technology.

**Table 7: Collaborative Institutions**

| Sl.No | Collaborative Institutions | No of Publication |
|---|---|---|
| 1. | Coimbatore Institute of Technology | 141 |
| 2. | Anna University | 97 |
| 3. | Kumaraguru College of Technology | 89 |
| 4. | Bannari Amman Institute of Technology | 83 |
| 5. | Bharathiar University | 58 |
| 6. | Indira Gandhi Centre for Atomic Research | 52 |
| 7. | National Institute of Technology Tiruchirappalli | 49 |
| 8. | PSG College of Arts and Science | 49 |
| 9. | Karpagam College of Engineering | 34 |
| 10. | Avinashilingam University for Women | 33 |

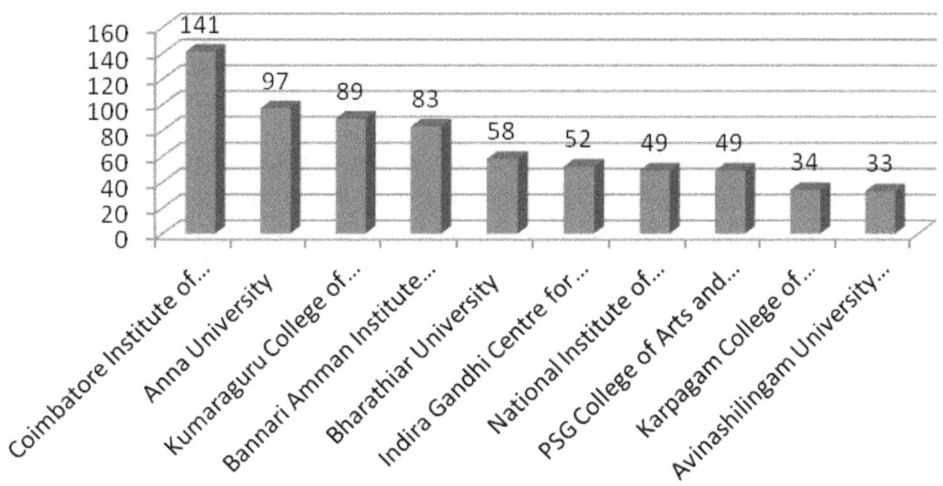


**Figure 6– Collaborative Institutions**

## 7. Discussions

Analysis shows that the faculties of PSG College of Technology has published highest number of publications in the year 2014 and also it is shows that the growth rate of publication has steeply increased during the period of study 2005-2015. In between 2008, slightly decreasing in output of research has been noticed. Document wise, total of 3425 papers published in article type and also fully published only in English language. The authorship pattern reveals that multi authored papers are more than single authored papers. Most of the papers were written by two authors. Since scientific research is a team work, generally the multi authored papers will be more than individual work.

## 8. Conclusion

This paper has highlighted quantitatively the contribution made by the faculties of PSG College of Technology during 1978-2016 as reflected in Scopus database. The scholarly production of papers at PSG College of Technology gradually increases in numbers;In the year 2014 highest number of articles has published with more number two authored articles by the faculty of PSG College of Technology. This study clearly shows that the publication trend is getting increased in the southern region of our country. There is a significant contribution of research article in the field of engineering and technology by the faculties and researchers collaboratively with the esteemed institutions.

## References

Anil Kumar& et.al.(2008). Pramana - Journal of Physics: A scientometric analysis,Annals of Library and Information Studies,55.Retrieved from http://eprints.rclis.org/bitstream/10760/14828/2/pramana.pdf on September 5, 2016

Arunachalam,S., (2001).Mathematical research in India today: what does the literature reveal?. Scientometrics,52(2) 235-259.

Dutt, B., Garg, K.C., & Bali, A. (2003). Scientometrics of the international journal Scientometrics. Scientometrics, 56(1), 81-93.

Hood, W.W. & Wilson, C. (2001).The literature of bibliometrics, scientometrics, andinformetrics. Scientometrics, 52 (2), 291–314.

Inzelt, A., Schubert, A., & Schubert, M. (2009). Incremental citation impact due to international co-authorship in Hungarian higher education institutions. Scientometrics, 78(1), 37–43.

Kavitha,M & Arumugam, J.(2011).Scientometric analysis of Indian contribution to mathematical research,Journal of Library, Information and Communication Technology,3(1-2), 61–70.

Mooghali,A., & et.al (2011).Scientometric analysis of the Scientometric literature, International Journal of Information Science and Management, 9(1). Retrieved from http://www.srlst.com/ijist/Vol9N1/ijism-V9N1_files/ijism91-19-31.pdf. on September 5, 2016

Nattar,S., (2009Indian journal of physics: A Scientometricanalysis,International Journal of Library and Information Science 1(4), 055-061.Retrieved fromhttp://www.academicjournals.org/ijlis/PDF/pdf2009/Sep/Nattar.pdf on September 5, 2016

Nikzad, M.(2012). Foreigners' point of view towards collaboration with Iranian authors. Webology, 9(2), Article 101. Retrieved September 5, 2016, from http://www.webology.org/2012/v9n2/a101.html

Rathinasabapathy, G., and L. Rajendran. "Mapping of world-wide camel research publications: A scientometric analysis." *Journal of Library, Information and Communication Technology* (2013).

Subramanyam, K. (1983). Bibliometric studies of research collaboration: a review. Journal of Information Science, 6, 35–37.

Wen et. al. (2007) Scientific Production of Electronic Health Record Research, 1991–2005" computer methods and programs in biomedicine, 86, 191–196

# 23

# Libraries Beyond Barriers: Assistive Technologies and Library Services to Physically Challenged Persons

**Amit Kumar**

Assistant Librarian
M.S.Randhawa Library, Punjab Agricultural University
Ludhiana, Punjab

**ABSTRACT**

*The Physically Challenged Persons constitute a very small percentage of the total population. The article presents the information regarding the infrastructure, the library services to the this special group of library users. All the public libraries do not provide services to these group of the users, hence the institutions specially meant for these persons are responsible for administering services to this group of users. Assistive information technology can help to a great extent in providing library services to these group of users.*

*Keywords: Physically Challenged Persons, Library Infrastructure, Library Services, Library Collections, Information Technology, Assistive Devices*

## 1. Introduction

The basic function of the library is to satisfy the intellectual needs of the readers. The libraries play important role in the lives of the Physically Challenged Persons (PCP) . Library is a social institution and must be planned and organized according to social needs i.e. the actual and potential needs the community. It is also expected to serve the community of special type of users i.e. PCP. In setting up a library for these users, certain basic factors are to be considered. It is apparently clear that these person are in need of library for various kinds of services, which are routinely available to the normal people, especially in himachal pradesh. They have depended largely on the voluntary organizations providing services to these people.

## 2. Physically Challenged Persons

Generally speaking the term "Physically Challenged" means disadvantage for a given individual, resulting from an impairment of disability that prevents of restricts the fulfillment of a role that is normal (depending on the sex and social cultural factors) for the individual. According to World Health Organisation a "Physically Challenged Person" is one who has "impairment, i.e. any loss or abnormality of psychological, physiological, anatomical structure of function." A Physically Challenged Person (PCP) lacks to perform an activity in manner or within the range considered normal for human being. In context of Libraries, a Physically Challenged client is one whose disability prevents him from performing activities performed by other users i.e. persons with orthopedic disability perform physical activity depending upon extent of disability, ones with speech and hearing disability face the problem of communication gap. The disability can be by birth or can be caused by accident or due to malnutrition and other unhygienic conditions.

Librarians possess a very important position not only in the libraries but also in the parent organizations as well. They are mediators between the users and the collection of the library. They are the ones who know what user wants and how the documents can be fully used

## 3. Objectives of the Study

1. Infrastucture for library services to PCP.
2. Library staff and collections for these libraries.
3. Assistive technologies used for providing library services.

## 4. Library Building

It is generally recognized that access to the library is a vital issue to be considered by the service providers to the PCP. The PCP shall be discouraged from using the library altogether if there are too many barriers to access the same. The library building would be accessible and would consider alternatives to service provisions for the PCP. Physical access and navigation within buildings require some modifications to help the PCP in accessing the information sources independently. So the library buildings should have modifications like proper signage, talking signage and elevators, ramps, etc. to provide the PCP with the physical access to the library building freely and conveniently. The libraries are generally should not be housed in a single room with no facilities. These should have proper lighting and ventilation therein. These libraries should not be located on the top floors and even in the basements. Proper signage should be in place to assist and give directions to the PCP. People with disabilities should be able to arrive at the site, approach the library building and enter the building conveniently and safely.

If the main entrance cannot be made accessible, an alternative accessible entrance, equipped with automatic door opener, a ramp, and a telephone should be provided. Sufficient parking spaces nearer to the entrance of the library, marked with the international symbol for the disabled should be provided/reserved for the Physically Challenged Persons who come to avail of the library services there. The displays, signposting, etc. should be clear and easy to read there. Unobstructed and well lighted access paths to the entrance smooth and non-slip surface at the entrance should be ensured. If needed, a non-slip and not too steep ramp with railings next to the stairs should be provided. Railings on both sides of the ramp and entry phone accessible for the deaf users are also very essential.

## 5. Getting into the Library

A person in a wheelchair or using crutches or a walker should be able to enter through the door and pass through security check points. There should be:

i. sufficient space in front of the door to allow a wheelchair to turn around ;

ii. entrance door wide enough to allow a wheelchair to enter;

iii. automatic door opener reachable by a person in a wheelchair;

iv. no doorsteps -- for easy wheelchair access;

v. security checkpoints possible to pass through with a wheelchair/walker or other mobility aides;

vi. stairs and steps marked with a contrasting colour;

vii. pictogram signs leading to the elevators;

viii. well lighted elevators with buttons and signs and having synthetic speech ; and

ix. elevator buttons reachable from a wheelchair.

## 6. Access to the Materials and Services

All parts of the library should be accessible. The space should be logically arranged with clear signs and a floor plan posted close to the entrance. Service desks should be located close to the entrance. Wheelchairs should be able to move around inside all parts/sections of the whole library. There should be no doorsteps and all doors should have automatic openers. Ideally, shelves should be reachable from a wheelchair. A certain number of tables and computer workstations should be adapted for persons in wheelchairs. There should be at least one toilet meant for the disabled persons.

## 7. The physical space

Special attention should be paid to ensure sufficient physical space in the library. In order to ensure this, the following steps should be taken:

i. clear and easy-to-read signs with pictograms;
ii. shelves reachable from a wheelchair;
iii. reading and computer tables of convenient heights throughout the library;
iv. chairs with sturdy arm rests;
v. unobstructed aisles between the bookcases;
vi. Visible and audible fire alarm; and
vii. Staff trained to assist the patrons in case of emergency.

## 8. Toilets (Washrooms/Restrooms)

The library should have at least one toilet for the disabled persons, equipped with the following:

i. clear signs with pictogram indicating the location of the toilet;
ii. door wide enough for a wheelchair to enter and sufficient space for a wheelchair to turn around;
iii. room enough for a wheelchair to pull up next to the toilet seat;
iv. Toilet with handles and flushing lever reachable for persons in wheelchairs;
v. alarm button reachable for persons in a wheelchairs;
vi. washbasin, mirror at the appropriate height, convenient for a person in a wheelchair.

## 9. Staff of the Libraries

The library staff plays a very important role in providing services to the PCP. The size and background of the staff assigned for the functioning of the library is an indicator of the quality of the services provided therein. The status of the staff is also indicator of the ability to provide an adequate range and quality of services and as well as an indicator of the level of organizational support for operation. The staff required in the libraries can be categorized as "Professional Staff" and "Supporting Staff". The "Supporting Staff" may also be called a "Semi-professional" or "Non-Professional" staff. The professional staff can further be categorized as Librarian, Deputy Librarian, Assistant Librarian and Junior Library Assistant. The libraries serving the PCP are different from those serving the normal persons. They differ not only in terms of users they serve but also in terms of services, equipments and reading material. To cope up with these changes, the professional staff of the library requires special training to provide specialized services to their clients. It is important that the entire library staff should have the understanding of the PCP to enable them to use equipment and aids designed to give the PCP access to the library's stock. The staff is required to be trained for providing services to the Physically Challenged Persons. Requirements of the users differ from person to person. The staff familiar with collections of the library, helps in fulfilling the needs of the PCP by selecting the material as per their requirements is known as "Reader's Service" which should be available in libraries serving the PCP.

## 10. Library Collection

A library's collection is the heart of its services and a reflection of its mission. Building a good collection to meet the needs of a community requires consideration of its age, economic, cultural and racial diversity. The PCP are equally diverse clientele with same reading needs as those of any other group. They require collection that includes popular magazines, information on course work, jobs, etc. the collection should support the reading needs of all ages of the disabled persons right from the children to the adults. The PCP have numerous and variety of information requirements from the library. They also need appropriate formats or aids to help them to access their required information. The alternative formats include e-books, audio cassettes or CD and other formats include assistive softwares. They need recreational reading, educational materials, encyclopedias, directories and all other kinds of documents, which are used by the normal people. The basic requirement of the PCP is the textbooks but they also like reading other materials like magazines, etc. Newspapers and magazines are current sources of information and help to keep one up to date. This is a very important kind of material. Although the magazines and newspapers are available in all the libraries, a very small percentage of libraries have these in alternative format i.e. in electronic form or audio format. The number of periodicals and newspapers subscribed by the libraries are the measures often used to compare the size of the library. Now a days these are available in alternative formats i.e. e-newspaper, talking newspapers, digital, etc.

With the advent of information technology, the access to information has become easier for the PCP and, therefore, it is essential that libraries offer the state of the art technologies for providing better library services to the PCP. There is a range of equipments, which can be used by the PCP for accessing information from the library. These range from photocopying machines to closed circuit television cameras capable of magnifying. They can read the text aloud and can employ scanner for generating a file for use with the help of the computer.

## 11. Library Services

The libraries play a very important role in providing services to the PCP. The library may be of any type i.e. public, academic, or special, the services and operations of all these are aimed at providing information to the users. The information is disseminated through the services provided by the libraries. Hence the libraries for the PCP are required to plan their services very carefully, taking into consideration the needs of these groups of users. The PCP can avail the library services through postal service, mobile, email, fax, etc. while sitting at their places. Postal and mobile services are important for house bound the PCP adults. All the library services must identify the needs, develop policies, earmark resources and plan services in such a way that the PCP have access to the same range of services as everyone else. This should be achieved by integrating the service requirements of the PCP into mainstream services. No library service is self sufficient, either in the range of subjects covered or in the books on its shelves. Therefore, Inter Library Lending between libraries has been a long established practice. The libraries function as a part of wider network providing better library services. These services help the PCP to access the information in desired formats.

The support services are required to help the PCP in using the collections of the library, which they are unable to access. The libraries should provide these support services to the PCP according to their disability. Online Library Catalogue is one of the important source of information for providing services to the users. The computer related aids and equipments are not easy to learn and hence they become barriers in accessing information by the PCP. The libraries should provide training to use assistive devices to the PCP. The library is a growing organism and the services to the PCP would be impossible without identifying and promoting "new arrivals" services. Different libraries adopt different methods for promoting these services, these services should be provided in according to the needs of PCP. The libraries have to select the reading material for developing the collection. The material should be selected on the basis of criteria, which should reflect the needs of the PCP community and should be balanced to meet their subject, recreational or information needs

## 11.1. Circulation Desk

Since the Physically Challenged Persons are required to get the documents issued for use at home also, the circulation desk should have the following facilities:

i. adjustable desk;

ii. induction loop system for hearing impaired persons;

iii. chairs for elderly and disabled persons;

iv. accessible self-service circulation stations

## 11.2. Reference/Information Desk

Reference/Information service is one of the important library services required by the library users. The Physically Challenged Persons are no exceptions to this service. Rather they need it much more, as compared to other library users. The following steps can be of great help for the Physically Challenged Persons in this regard:

i. adjustable desk;

ii. organized "queue system" in the waiting area;

iii. chairs suitable for the elderly and disabled patrons;

iv. induction of loop system for the hearing impaired persons.

## 11.3. Persons with Reading, Hearing and other Disabilities

Patrons with reading disabilities need special attention when they visit the library. The library staff should be knowledgeable about various disabilities and how to serve patrons with these disabilities. Materials specifically produced for the persons with reading disabilities should be easy to find. The following steps should be taken to help the library users having reading, hearing and other physical disabilities:

i. a centrally located department with talking books and other materials for the persons with reading disabilities should be established/earmarked;

ii. a colored (yellow for visibility) tactile line leading to this special

iii. department should be provided;

iv. clear signs should be displayed;

v. comfortable seating area with bright reading light should be provided;

## 11.4. Computers

Computers for public use should be accessible. Fast and reliable technical support should be available for both computers and adaptive equipment. The library staff should be trained to provide on-site support. Necessary steps should be taken to ensure the following facilities for the Physically Challenged Persons:

i. designated computer workstations for the patrons in wheelchairs;

ii. adaptive keyboards or keyboard overlays for users with motor impairments;

iii. designated computers equipped with screen reading programs, enlargement and synthetic speech;

iv. designated computers equipped with spelling and other instructional software suitable for persons with dyslexia;

v. technical support for computers (on-site, if possible); and

vi. staff capable of instructing users in the use of computers.

vii. a tape recorder, CD player, Digital Audio Information System (DAISY) and other equipment to complement the audiovisual collection should be made available;

viii. magnifying glass, illuminated magnifier, electronic reader or Closed-Circuit Television (CCTV) should be provided; and

ix. computers with screen adapters and software designed for persons with reading and cognitive disabilities, should also be made available for the Physically Challenged Persons.

## 11.5. Telecommunication Devices for the Deaf (TDD)

The text telephone is a communication device which can be attached to the telephone to send typed messages back and forth over telephone lines to other individual with TDDs. The typical TDD is a device about the size of a typewriter or laptop computer with a QWERTY keyboard and small screen that uses LED or an LCD screen to display typed text electronically. In addition, TDDs commonly have a small spool of paper on which text is also printed — old versions of the device had only a printer and no screen. The text is transmitted live, via a telephone line, to a compatible device, i.e. one that uses a similar communication protocol. If a deaf person communicates with a person who does not have a TDD, he/she must call the operator who will place the call to a hearing person with a TDD who acts as a relayer.

## 11.6. Tele-typewriters (TTYs)

TTYs are devices that enable Deaf, Hard of Hearing, and Speech Impaired individuals to communicate with each other and hearing people via the telephone system. Similar in appearance to an electronic typewriter with a display screen, TTYs transmit and receive electronic signals that the receiving TTY converts to text and prints it on the TTY display screen. The display screen size varies from about 20 to 40 characters or more, depending on the TTY.

If you're a hearing person, but don't have a TTY and need to talk to someone who does, you can use the Relay Service to call them, or, in some cases, use your computer to call the individual. Some TTYs can communicate with computers (if the proper transmission mode is used and the computer user has a modem).

## 11.7. Captioning

Captioning is the process for converting audio information into text and displaying the text on a screen or monitor. It provides important visual information to millions of people who are Deaf or hard of hearing, including access to television programs, film, CD/DVDs, and live events. It can also serve to assist with improved fluency in the English language. It enables the hearing impaired to make and receive phone calls while using their own voice. Whoever they are conversing with has everything they say captioned on the screen. Calls are made in the same way as a normal telephone call. It is useful for Deaf and hearing impaired people: Who find it difficult to hear and understand telephone conversations, Who are able to talk clearly and read captions/

## 11.8. i Communicator

i Communicator is an amazing tool that enables people with hearing disabilities to effectively communicate. iCommunicator can act as alternative when an interpreter is unavailable, for example classroom settings, daily communication, ad hoc meetings, emergency situations etc. iCommunicator translates in real-time: Speech to Text, Speech/Text to Video Sign-Language Speech/Text to Computer Generated Voice.

## 11.9. Digital ALD's

Digital assistive listening systems are the newest products on the market. They are completely wireless so they are unobtrusive. They connect directly to digital hearing aids, and their quality is considered superior. Hearing aid users who want to use a digital system need to see their audiologist to be fitted for the correct "boot" which attaches to the hearing aid.

## 11.10. Remote Video Conference Sign Language Interpreting using Skype

Many people in the Deaf community who use sign language can access sign language interpreters through the internet using software such as Skype or OVOO. All that is required is a laptop or desktop computer with a fast internet connection . This is useful, particularly for people who are Deaf who reside in rural areas where access to sign language interpreters is limited. All that is required is a good quality webcam based computer. **'Skype'** – free to download, this benefits those with a hearing impairment, or those that rely on visual face-to-face communication. Skype is a form of video-conferencing using a webcam on the computer, laptop, TV, iPad, iPhone or iPod Touch with WiFi or 3g / 4G access. It can also be used for instant messaging.

## 11.11. LOMAK

(Light Operated Mouse And Keyboard) enables people with physical impairments, such as cerebral palsy, quadriplegia and carpal tunnel syndrome, to easily and effectively operate a computer. Using a specially designed keyboard in conjunction with state-of-the-art light sensor technology, a hand or head pointer controls a beam of light that enters, then confirms, the key or mouse function. Confirming each key helps ensure the correct selection is entered, reducing errors and increasing the speed of operation.

## 11.12. Adaptive keyboard

There are a wide range of alternative keyboards on the market to help motor-impaired users including compact, expanded, ergonomic, on-screen, concept, rubber and ABC keyboards.

## 11.13. Touchscreens

A touchscreen is a computer display screen that is sensitive to human touch, allowing a user to interact with the computer by touching an active area, target or control such as pictures or words on the screen. Touchscreens are activated by the insertion or removal of the fingertip or by pressing the controls active areas or targets with a mouthstick, headstick, or other similar device (stylus).

## 11.14. Voice recognition software

Voice recognition programs enable the user to enter text and, in some cases, carry out common computer tasks simply by speaking into a microphone – that is, without having to use a keyboard or a mouse. In case of text entry, the computer analyses the user's voice, tries to recognize the words, and types them - instead of the user - as he or she speaks. Using voice recognition software, the user can bypass the keyboard and just speak to the computer. By programming the computer with a set of predefined instructions, the user can control the computer by verbally issuing commands into a microphone.

### 11.15. Virtual mouse

A virtual mouse is a software program, a graphical on-screen mouse, that includes all the functionalities of a standard mouse and that can be used for example with a switch.

### 11.16. Single Purpose Assistive Listening Systems

A Pocket Talker is a personal listening system that can be helpful in one-on-one situations. A wire connects the transmitter to the receiver so the speaker must be positioned close to the listener and neither can move around. A Hearing Aid Telephone Interconnect System (HATIS) is an assistive listening device for a telephone. One part plugs into the audio jack of the phone, and the other part is placed over a hearing aid.

### 11.17. Assistance from the Government

Library is a growing organism and no library is self- sufficient. It depends on the assistance from various sources for providing appropriate services to its users. The libraries serving the PCP require more help as they are facing paucity of funds as well as scarcity of the skilled manpower. The libraries can improve the quality of library services to the PCP with the help of Government as well as Non Government Organizations. The Government can help the libraries in two ways, firstly, by laying down standards for providing services and issuing instructions to the concerned libraries and secondly by providing timely grants for improving the financial health of libraries.

## 12. Conclusion

Information and Communication Technology has been identified as an important aspect of the wider strategy for the social inclusion of the disabled people. ICT is heralded as enabling Disabled people to participate fully in social and economic life of their communities. ICT is a significant force in terms of choice and opportunity for the Disabled people. ICT offers the old and the young alike an opportunity to overcome social barriers to interaction and communication that can be caused by the lack of provision for impairments or life long limiting illness. ICT has also been identified as playing a significant role in offering severely disabled people an increased degree of independence in everyday life. ICT gives the Disabled person an improved quality of individuals the ability to compensate for physical or functional limitations, thus allowing them to enhance their social and economic integration in communities by enlarging the scope of activities available to them. ICT have revolutionized life of disabled people, they are used by deaf or hearing impaired, physically disabled, mentally handicapped.

The problems faced by the Library Authorities in running the libraries for PCP like paucity of funds, lack of staff, etc. are of trivial nature and can be overcome with little interest, enthusiasm and motivation on the part of the Ministry of Human Resource Development and Social Welfare Department. If the government has "written policies" for providing library services to the PCP, like In the countries like United States of America, the Prat Smoot Act of Congress established library services for the PCP in 1931. Although it is a department within Library of Congress, this service was recognized and legislated separately. This type of acts in India will definitely motivate the Library managements to upgrade their libraries in accordance with the standards proposed by the Government. Regular funds from the Government will also enhance the quality of library services to the PCP. These steps will help in establishing a well equipped library in every institution meant for the Physically Challenged Persons.

## References


Bae, K.J., Jeong, Y.S., Shim, W.S., Kwak, S.J. (2007). The ubiquitous library for blind and physically handicapped: a case study of LG Sangnam Library, Korea. *IFLA Journal, 33(3), 210–19.*

*Cylke, F. K., Moodie, M. M.; Fistick, R. E. (2007). Serving the blind and the physically handicapped in the United States of America. Library Trends, 55 (4),796–808.*

*Fischer, A. (2005). National library service for the blind and physically handicapped. Library of Congress Information Bulletin, 64(1),10.*

*Journal, 103(7), 725.*

Koulikordi, Anna ( 2008), Library services for people with disabilities in Greece, Library Review, 57(2).pp.138–148.

Lovejoy, E. (1978). Library services to the blind and physically handicapped. *Library*

*Needham, W.L. (1978). Library services to the blind and physically handicapped. Journal of Academic Librarianship, 4 (4), 225.*

Panella, Nancy. (2009). The library services to the people with special needs section of IFLA: An historical overview. IFLA Journal, 35(3).pp.258–71.

Qiu Fengjie and Wang Zizhou, (2009). A questionnaire survey on reading and public library utilization of the disabled in Beijing, 2009,CJLIS, 2(3), 62–81

Seth, M.K., Parida, B. (2006). Information needs and use patterns of disadvantaged communities : A case Study. *Library Philosophy and Practice*, 9 (1), 14–17.

Stout, A. (2006). Accessibility for the disabled at the New Zealand public libraries. *New Zealand Library and Information Management Journal*, 50(1),44–58.

Todaro, A.J. (2005). Library services for people with disabilities in Argentina. *New library World*, 106(5/6), 253–268.

# 24

# Collection Development in University Libraries in Epoch of Information Technology

**Dr. Rajive K. Pateria**

Deputy Librarian Nehru Library,
CCS Haryana Agricultural University, Hisar (India)
E-mail: rajivepateria@gmail.com

and

**Dr. Seema Parmar**

Assistant Librarian, Nehru Library,
CCS Haryana Agricultural University, Hisar (India)
E-mail: seemaparmar9@gmail.com

**Abstarat**

*Under the umbrella of Information and Communication Technology (ICT) Information resources in both print and electronic media are exploding in a wide range of academic libraries throughout the world. This detonation of resources leads to development of collection in academic libraries but this development of collection need some base on terms quality and quantity. The paper describes the components required for collection development in all university libraries so that an adequate and balanced collection could be developed to meet the ever changing needs and demands users and collection of any academic library and could satisfy teaching, research and other multifarious needs of the students and faculty members and help them to keep abreast of new developments in their field.*

## 1. Introduction

Libraries of all kinds are developing their collection under the shade of electronic environment to satisfy the all kind of informational thirst of their users, whose expectations are increasing day by day and their attitude towards information is gradually shifting from the printed documents to electronic resources. Librarians are now becoming the champions of information. Now, they have shown their interest not only in accepting the traditional printed documents for their libraries but also making new technology based library resources the part of their libraries' collection. Its main effect can be seen in development of e-collection in libraries. The collection growing in libraries in the present era can be classified as presented through Figure.1

## 2. Collection Development

"The terms connected with the collection development are not well defined in LIS Literature. The Collection development, selection, acquisition, collection building and collection management have all been used interchangeably" (Maheshwarappa, 1997, 155).

Many authors have defined the term 'Collection Development' in their own ways. According to Sachez Vignau (quoted in Sanchez Vignau and Meneses 2005, 35) Collection development is "a process which assumes that the information needs of the users are satisfied in economic fashion and inside of a reasonable period of time using resources as much internal as external to the organization".

On the other hand, Johnson and Osburn (quoted in Walmiki, Ramkrishnegowda, Padamma and Prithviraj 2012, 9) has given a very functional definition of collection development as he defines "collection development cover several activities including the determination and coordination of selection policy, assessment of the needs of users, collection analysis, selection, budget management and planning for resources sharing".

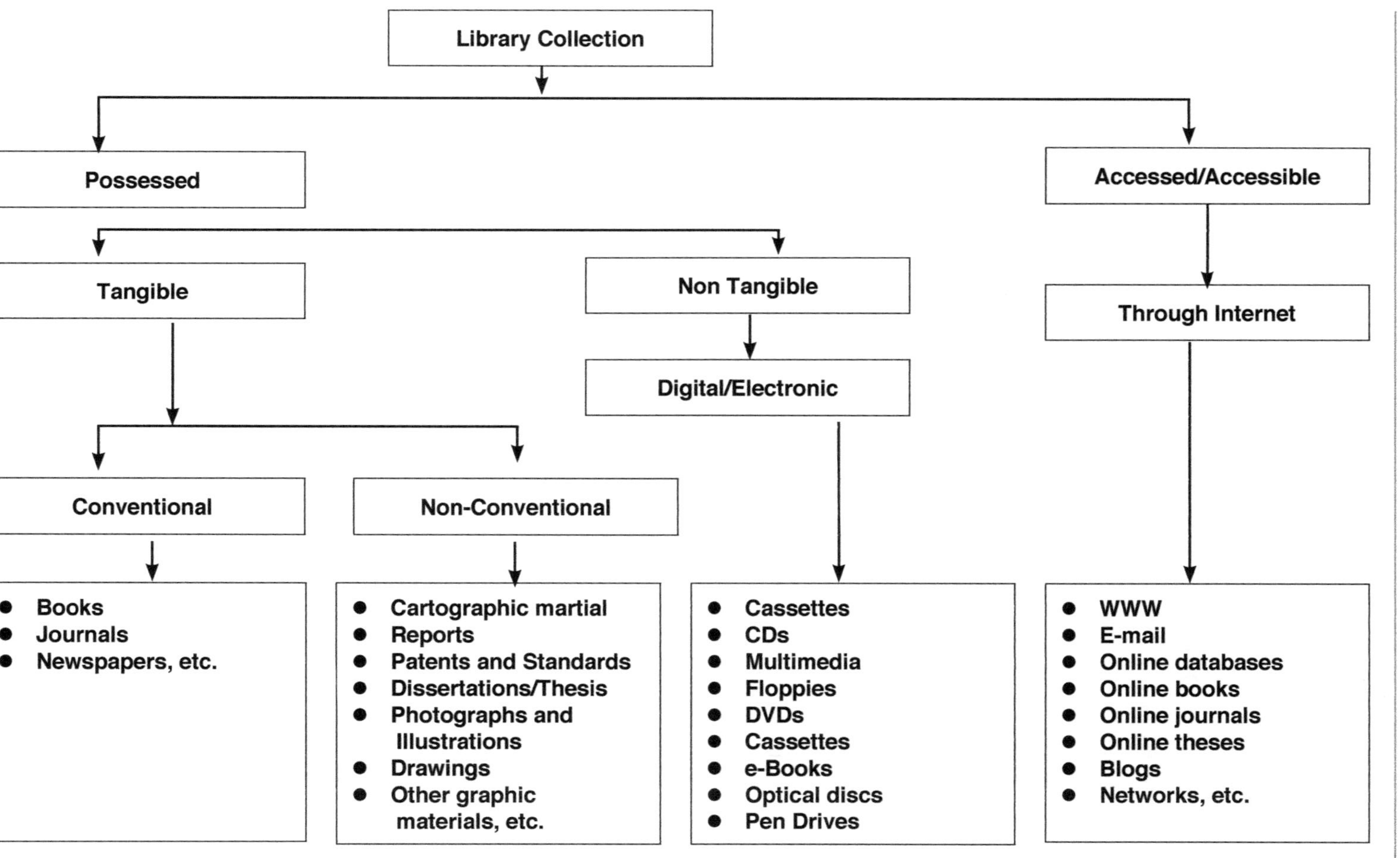


Fig. 1 Classification of library collection in I.T era

In brief we can say that collection development embrace regular assessment of the informational needs of users, uses statistics, demographic projection of user community, financial constraints, determining of selection and acquisition criteria, planning for resource sharing, evaluation of collection and weeding out unrequired items or reading material.

## 3. Collection Development in University Libraries in I.T. ERA

As the quality of teaching and research of a university depends upon the quality of information services provided by its library to its user community, the aim of a university library should be to build an adequate and balanced collection to meet the ever changing needs and demands of the target audience. So that, the collection of library could satisfy teaching, research and other multifarious needs of the students and faculty members and help them to keep abreast of new developments in their field. Thus, collection development of a university library should be need based in term of quality and quantity.

Collection development in the present dynamic environment is a challenge for every university library as various "issues like interdisciplinary nature of research, information explosion, production and availability of information sources, multimedia, automation of library system, physical deterioration of documentary resources, changing concept of ownership to access, library networks, internet services and financial constraints" effects their activities and services (Maheswarappa and Tadasad, 1997, 25).

## 4. Components of Collection Development

Many authors have differently described the components of Collection development. Some authors have explained them as functions of collection development; some elaborated them as elements of collection development and other few have considered them as processes of collection development.

Collection development can be cluster of different components which collectively takes the shape of whole holdings of any library. These components are:

- Users' analysis
- Selection
- Acquisition
- Resource sharing
- Collection evaluation
- Weeding

## 5. Users' Analysis

Users' analysis or community analysis is first function of collection development. For collection development in any library, it is essential to know which type of information the user community needs. University libraries have different category of users such as academic, administrative, technical, professional, researchers, students and so on. Information needs of the users vary according to their academic status and assignment on which they are engaged at the given time. Librarian should conduct a sort of user studies to know about the changing information needs of different category of users to meet their information requirement adequately. Collection of the university library must be developed in the way that suit to its users' community.

Users' inputs collected through the process of users' analysis provides information about reading habits and interest area of the users, strength and weakness of the existing collection, suggestion for improvement of collection, and preferred formats of resources. These inputs help the librarians to develop their libraries' collection in an effective manner.

## 6. Selection

"Selection process begins with determining the type of materials to be added such as format, genre, subject, language, geographical coverage in the light of user community on the one hand, and mission, goals and priorities of library and its parent institution on the other" (Walmiki, Ramkrishnegowda, Padamma and Prithviraj 2012, 11). The selection of resources in a library involves various processes like identification of resources, assessment and evaluation and decision to purchase. Selection of material is considered the joint responsibility of users and library staff. Some university libraries constitute selection committees for this purpose consisting faculty, student and library staff. University libraries acquire different types of information sources in different formats to meet the information needs of different categories of users. In order to develop quality collection, libraries should frame some policy or guidelines for selection of material to be added to help users and library staff to take decision in selection.

Some important criteria used for selection of print material to be added in library are their contents, currency, price, reputation and authoritativeness of the author, publisher and suitability as per the need of the library. Traditional selection criteria are still valid while selecting e-resources for the library but these resources require more exhaustive criteria than print resources

**Technical issues:** Only those libraries can think of selection of e-resources, who have applicable technology to access them. Various factors are necessary to be considered while selecting e-resources for the collection of library and they are: a) number of technology acquainted staff in the library and number of staff required training; b) compatibility of formats of e-resources with the existing hardware and software; c) technical support given by vendors d) browser for searching; e) various formats of digital access (CD-ROMs, remote access etc.); f) possibility of integration with other e-resources g) access management, etc. All theses require depth knowledge of computer and network technologies.

**Legal issues**: According Trisha L Davis "Licensing agreements have become a fact of life in the electronic publishing world. Yet nothing about them is standard or predictable" (quoted in Wilkinson 2005, 142). Licensing agreements may often include provisions for pricing, payment and delivery of the product, limit access, define use, warranting and limits, termination of the agreement, customer service, quality and currency of information and responsibility of the license for the security of the product. While selecting e-resources for the library a selector should consult with vendors and aggregators to assure that all the legal needs are met before the selection decision is made. The library should carefully review licensing terms before the purchase of e-resources and identify what clauses should be included in the agreements and what clauses are unacceptable.

**Pricing issues:** Different types of e-resources like e-books, e-journals, bibliographic databases, fulltext databases are published on different media like CDs, Online, etc. and their subscriptions are available through various agencies like publishers, vendors, aggregators, etc. Prices of e-resources for individual, institutional and consortia subscription vary depending upon type of access. Thus, Pricing issue of e-resources is also a challenging issue. Prices are different as per the type of subscription of the libraries like- different prices are taken from those libraries who subscribe only e-version and different for those who subscribe print with e-version and for those subscribers who subscribe only a few titles out of the bundle of resources. Therefore, at the time of selection of e-resources librarians should survey the different pricing models and get the information from other libraries where related e-resources have already been subscribed.

In case of web based e-journals some of the prevailing pricing models available are: a) electronic subscription is linked to the print subscription; b) electronic subscription with campus licenses; c) electronic subscriptions are bundled; d) paper-per-look; e) electronic only; f) consortium licensing; g) national licensing ( Arora, 2001).

Thus, selection of e-resources is a complex job for librarians due to involvement of these technical, legal and pricing issues. Nowadays Internet has become a powerful selection tool as most of information about latest published print and electronic resources, publishers' catalogues, vendors, stock list of book distributors, online library catalogues, book reviews, search engines are available on Internet and updated regularly. In this way it has not only made the selection process easier and faster but also saves money spent on purchase of costly selection tools.

## 7. Acquisition

Once the selection process of collection for the library has been completed, another activity to subscribe the selected resources starts in accordance with the resolute terms and trades. The purchase of print resources "proceeds with preparation of an order form, selection of a vender, the recording of the receipt of the item, and finally, payment after verifying the items" (Faruqi, 1997, 18). The process of acquisition of electronic resources is almost same as of print material as it also includes identification, order, receiving and payment for e-resources, but it is more complex and time consuming than the process of acquiring other print material. E-resources are complex in nature due to their various types, access models, subscription modes, technological involvement and licensing agreements. To acquire e-resources librarian have to perform several activities like identification of resources, reviewing licensing agreement, identifying cost, identifying level of access, technical support, vender selection, placing order, signing agreement, receiving / beginning access, paying and maintaining access.

The acquisition programme should be organized and implemented systematically and scientifically so that the reading material of maximum utility can be acquired with minimum cost in least possible time. "The success of implementing the selection and acquisition policies mainly depends upon two major factors- a) budgeting and b) staffing" (Jagannathan 1999, 193 &194).

### 7.1. Gift/Exchange

The libraries receive bulk of gifted material. Most of the material received in this manner is worthless and is of no interest of library. Libraries cannot acquire all the material received in gratis. "The golden rule for gifts is: Do not add a gift unless it is something the library would buy. Selectors must resist the temptation to add an item because it is free. No donated item is ever "free". Processing costs are the same for gifts and purchased material" (Evans 2004, pg. 82). Thus gifted items should be added to library collection after evaluation on the criteria already decided for selection of material.

### 7.2. Budgeting

The budget while purchasing information sources for library should be carefully planned and allocated keeping a few things in mind at the time of allocation: Number of departments; Number of certificate, diploma, under graduate and post graduate courses; Number of teachers and scientists; Number of students; Number of patrons and outside members; and Variation in prices of information resources in different subject areas. A formula for budget allocation may be designed considering the above criteria. Expenditure on purchase of materials should be in accordance to budget allocated to each department/subject during the current financial year. Thus the budget for library should be planned in such a manner that it may have provision to develop balance collection, both traditional and electronic to meet the information needs of all users groups. It should include estimated cost of reading material to be purchased, processing cost material and cost of computers infrastructure required for providing access to the user community.

### 7.3. Staffing

For the whole process of collection development like selection, acquisition, resource sharing, evaluation and weeding out of useless materials, it is essential to have plentiful, qualified, devoted and experienced staff so that to utilize the funds judiciously; to purchase required material by community; to process collection technically; to evaluate the stock; and to weed out the outdated and unused material or store them separated in a compact storage. While acquiring of e-resources, the staff should have depth knowledge of computer and network technology, information industry, legal matters, and capability to update themselves with the advancement of technology and latest market trends of information industry.

## 8. Resource Sharing

Most of the libraries in India are confronting the problem of insufficient finance and inadequate collection, thus, feel helpless to acquire all the documents published in different formats over the globe so that to meet the increasing informational needs of their users. To improve the situation, only one solution seems to be feasible and that is to practice of resource sharing.

Resource sharing is one of the important functions of collection development. It is considered as means to optimize the accessibility of material and reduce the cost without sacrificing identity. Libraries perform resource sharing activities through various ways like signing agreements with other libraries and seeking membership of networks and consortia etc. at local, national and international level. Some resource sharing activities performed by the libraries for collection development are:

Cooperative collection development: In this activity each participating library takes primary responsibility to develop collection in specific area or discipline and share the collection with other libraries free of charge.

Coordinated Acquisition: In coordinated acquisition, participating libraries agree to purchase certain reading material on cost sharing basis and one or more of the member libraries house the reading material.

Joint acquisition: In joint acquisition, one of the member library place joint order on the behalf of all libraries for purchase of reading material and each member library receives the material.

Inter Library Loan (ILL): In this activity member libraries provide requested to print the reading material to one another on loan basis for a specific period.

Document Delivery Service (DDS): DDS is a service whereby member libraries provide reprints or electronic copy of the requested document to one another.

Reciprocal borrowing privileges to patrons of member libraries

In these ways libraries may get the maximum discount on the purchases of reading material and save the cost on acquisition of reading material and provide better resources and services to their patrons within limited resources. Accessibility to member libraries' OPACs or union catalogue may help in effective implementation of resource sharing programme. Advancement of network and web technology has made resource sharing easy and faster. Now libraries are forming network and consortia for accessing e-resources and providing effective services to their patrons using network and internet.

### 8.1. Library consortia

Nowadays the term 'library consortia' is commonly known for sharing of e-resource. Most of academic and special libraries in India are joining the e-resource consortia to get access of e-resources at cheaper cost as "consortia provides union strength to negotiate with electronic publishers for best possible price and rights" (Arora 2001, 22). Thus, Library consortium is cooperative project to provide access to a wide range of e-resources to member libraries at least cost. The major consortia projects in the field of Agricultural Sciences are: CeRA, Krishipraha and e-Granth.

## 9. Collection Evaluation

The process of collection Evaluation is considered as the last component of collection development which avoids unwanted and irrelevant resources in the library. It is an assessment to know that how well the existing resources of the library meet up the users' informational requirements.

Evaluation of the digital resources of the library can be conducted by considering the following criteria: a) Amount spent on subscription of electronic resources; b) number of users accessed in-housed, networked and online databases/resources; c) number of documents downloaded from the online databases; and d) number of users' login to use the online databases.Systemically accumulated information on the scope, usefulness, accessibility and quality collection helps to evaluate the print as well as electronic resources.

## 10. Weeding

Weeding is the process of discarding or transferring excess copies, rarely used books and materials of no longer use. It is shifting of books from active shelves to other location of less importance and cost. It is necessary that the process of weeding or discarding should be carried out regularly, otherwise, it might give a big jolt in collection development and ultimately, the collection would become full of irrelevant and out dated . Therefore, every university library has to conduct evaluative study of its collection time to time and remove the unwanted and dead stock which is outdated or not being used to make space for new additions. Before the weeding out process is undertaken, well judged decisions regarding the type of materials to be discarded should be finalised.

## 11. Collection Development Policy (CDP)

Keeping in view of the limited funds available, increasing informational needs of user community and several handicaps existing, the collection development in libraries must be carefully and systematically planned in accordance with the mission, objective, scope of university programmes. This planning in the form of policy is an important element for print as well as digital collection building.

CDP should be framed with considering the mission and aims of individual library to meet the current and future needs of its users. It should be a living document, adoptable to change and grow. The guidelines provided in the policy document can be modified, as the collection needs change.

**Benefits of CDP for library:** There are following benefits of CDPs for libraries:

- It helps the library staff to understand the goals of the library and make them committed towards these goals.
- It helps staff to identify ever changing , ever increasing and different type of needs of users as well as the nature and scope of the collection so that they could think and decide the priorities while allocating funds on different type of collection.

- It helps in establishing standards for selection and weeding of reading material.
- It helps in promoting other libraries to facilitate coordination for collection development.
- It assists in minimizing influence of a single selector and personal biasness.
- It serves like a training tool for in-service and new appointed staff.
- It helps in evaluation of collection by staff themselves or for evaluation by outsiders..
- It contributes to enhance the operational efficiency in terms of taking day to day decisions
- It serves as a tool of complaint handling with regards to inclusion or exclusion.

### 11.1. Elements of CDP

A policy document should have the clear statement of – a) short and long term objectives for the library regarding the user community their informational needs and parameters of the library collection; b) status of collection in subject areas specifying the main user group for each subject ; c) selection criteria according the media like print, online, CD-ROM etc; d) responsibility of selection; e) preference between print and electronic resources; f) acquisition procedures; and g) policy revision.

## 12. Collection Development Policy in Indian Libraries

A number of efforts have been made by university libraries to develop CDP at international as well as national level but only a few foreign countries have attained the target while in India it has not came in to practice. The main reason is that the electronic collection in every Indian university libraries is haphazard till now. "So, far written collection development policies have found much less favour in Indian academic libraries, though they have their own internal policy developed by the Library Advisory Committee (LAC)" (Varalakshmi 2004, 86). Gandhi, Ramesh R.T.D (2001) in his study revealed that six libraries of Karnataka State did not had their well defined written collection development policies but they had Library Advisory Committee under the chairmanship of Vice-Chancellor which, time to time, laid guidelines on functioning of libraries and their development including selection, acquisition and collection of library material. Mal, Bidyut Kumar (2011) explored in his study after a period of ten years that none of twelve university libraries of state of Uttar Pradesh had written policy document and suggested to frame such policy as it is essential to have it for every library.

## References

Arora, Jagdish. "Electronic Publishing: an overview." *Joint Workshop on Digital Libraries United States Educational Foundation in India,* DRTC / Indian Statistical Institute, DLIS – Univ. of Mysore, Mar. 12-16 (2001). Accessed Sept. 11, 2011. http://isibang.ac.in/bitstream/handle/1849/128/Arora1.pdf?sequence=2.

Evans, G Edward. *Developing Library and Information Centre Collections, 4th ed.* Colorado: Libraries Unlimited, 2004.

Faruqui, Khalid Kamal, ed. *Development of Collection in Libraries.* New Delhi: Anmol Publications, 1997.

Gandhi, Ramesh R T D. "A study of problems and prospects of libraries and publishers with special reference to collection development in University libraries in Karnataka." Doctoral theses, University of Mysore, 2001. Accessed Dec.25, 2010. http://dspace.vidyanidhi.org.in:8080/dspace/ bitstream/2009/2934/6/UOM-2001-1576-5.pdf

Gessesse, Kebede. Collection development and management in the twenty-first century with special reference to academic libraries: an overview. *Library Management* 21, no. 7 (2000): 365–372.

Jagannathan, Neela. "Collection management in academic libraries." *Library and Information Services in the Electronic Information Era: Seminar Papers* [44th All India Library Conference hosted by P S Telugu University, Hyderabad, 25-28 December, 1999], edited by J. L. Sardana, 188-196. Delhi: ILA, 1999.

Maheshwarappa, B. S. "Problems of collection development among academic libraries." *Herald of Library Science* 36, no. 3–4 (1997): 155–160.

Maheswarappa, B.S. and P. C. Tadasad. "Collection development in the context of electronic publications and networking: problems and prospects." *DESIDOC Bulletin of Information Techn*ology 17, no. 1 (1997): 25–31.

Marwade, Rajendra M. *Information Technology and Library Development.* ABD Publishers: Jaipur, 2012.

Marry, A. Lawrence and A. Shankar. Collection evaluation of PSN College of Engineering and Technology Library and PET Engineering College Library in Tirunelveli District. *SERLS Journal of Information Management* 45, no. 1(2008): 63–70.

Mohsenzadeh, Farnak and Isfandyari-Moghaddam, Alireza. "Application of information technology in academic libraries." Electronic Library 27, no. 6 (2009) : 986-998. Accessed July 19, 2012. doi: 10.1108/02640470911004075.

The New Oxford American Dictionary, New York: Oxford University Press, 2001.

Ugah, Akobundu Dike. "Size and Quality of Information Sources and the Use of Library Services in a University Library." *Library Philosophy and Practice* (2011). Accesses Feb., 2012. http://www.webpages.uidaho.edu/~mbolin/ ugah8.htm

Van Zijl, Carol. "The why, what, and how of collection development policies?" *South African Journal of Library and Information Science* 66, no. 3 (1998): 99–106.

Varalakshmi, R. S. R. "Collection management in digital environment: policy statement for university libraries". SERLS Journal of Information Management 41, no.1 (2004): 79-90. Sanchez Vignau, Barbara Susana and Grizly Menses. Collection development policies in university libraries: A space for reflection. *Collection Building* 24, no.1 (2005): 35-43. Accessed April 11, 2011.doi: 10.1108/01604950510576119.

Walmiki, R. H., Ramkrishnegowda, K. C., Padamma, S. and Prithviraj, K.R. "Collection development in electronic era: some issues." *Collection Development in the Electronic Environment* [Proceeding of the International Conference on Collection Development in the Electronic Environment, Madras University Library, Madras, 29-30 June, 2012], edited B. Ramesh Babu, K. Kaliyaperumal, P. Rajendran and S. K. Asok Kumar, 9-18. Madras: Madras University Library, 2012.

Wilkinson, Frances C and Linda K Lewis. *The Complete Guide to Acquisition Management.* Libraries Unlimited: Westport, 2005.

www.ingramcontent.com/pod-product-compliance
Ingram Content Group UK Ltd.
Pitfield, Milton Keynes, MK11 3LW, UK
UKHW021532300726
14060UKWH00011B/366

9 789389 605228